ASTRONOMY AND
ASTROPHYSICS LIBRARY

Series Editors: M. Harwit, R. Kippenhahn, V. Trimble, J.-P. Zahn

ASTRONOMY AND
ASTROPHYSICS LIBRARY

Series Editors: M. Harwit, R. Kippenhahn, V. Trimble, J.-P. Zahn

Michael H. Soffel

Relativity in Astrometry, Celestial Mechanics and Geodesy

With 32 Figures

Springer-Verlag Berlin Heidelberg New York
London Paris Tokyo

Privatdozent Dr. M. H. Soffel

Lehrstuhl für Theoretische Astrophysik, Universität Tübingen
Auf der Morgenstelle 12, D-7400 Tübingen, FRG

Series Editors

Martin Harwit

The National Air and Space
Museum, Smithsonian Institution
7th St. and Independence Ave. S.W.
Washington, DC 20560, USA

Virginia Trimble

Astronomy Program
University of Maryland
College Park, MD 20742, USA
and Department of Physics
University of California
Irvine, CA 92717, USA

Rudolf Kippenhahn

Max-Planck-Institut für
Physik und Astrophysik
Institut für Astrophysik
Karl-Schwarzschild-Straße 1
D-8046 Garching, Fed. Rep. of Germany

Jean-Paul Zahn

Université Paul Sabatier
Observatoires du Pic-du-Midi
et de Toulouse
14, Avenue Edouard-Belin
F-31400 Toulouse, France

Cover picture: Astrometry, Celestial Mechanics and Geodesy: among modern space techniques
are radio interferometry with signals from extragalactic radio sources, laser ranging to retro-
reflectors on the lunar surface or satellite laser ranging to the LAser GEOdynamical Satellite
LAGEOS.

ISBN 3-642-73408-1 Springer-Verlag Berlin Heidelberg New York

Library of Congress Cataloging-in-Publication Data. Soffel, Michael H., 1953– Relativity in astronomy,
celestial mechanics, and geodesy / Michael H. Soffel. p. cm. — (Astronomy and astrophysics library)
Bibliography: p. Includes index. ISBN 0-387-18906-8 (alk. paper) 1. Astrometry. 2. Mechanics, Celestial.
3. Geodesy. 4. General relativity (Physics) I. Title. II. Series. QB807.S64 1989 521.1—dc19 88-37327

2156/3150-543210 – Printed on acid-free paper

Foreword

The book "Relativity in Astrometry, Celestial Mechanics and Geodesy" represents a significant contribution to modern relativistic celestial mechanics and astrometry. In these branches of astronomy the theory of general relativity is used nowadays as an efficient practical framework for constructing accurate dynamical theories of motion of celestial bodies and discussing high-precision observations. The author develops the useful tools for this purpose and introduces the reader into the modern state of the art in these domains.

More specifically, the distinctive feature of the book is the wide application of the tetrad formalism to astronomical problems. One may not agree with the author's opinion that this is the only method so far to be able to treat the relativistic astronomical problems in a consistent and satisfactory manner. (On the contrary, one may foresee in the nearest future other books on relativistic celestial mechanics and astrometry based on different approaches solving the same problems.) However, we are now at the beginning of practical relativistic astronomy and it will demand much effort to reconstruct in a relativistic manner all Newtonian conceptions of ephemeris astronomy and geodesy. In particular, this concerns the definitions of reference frames, time scales and astronomical units of measurement. This book is one of the first steps in the correct direction.

V.A. Brumberg, Institute of Applied Astronomy, 196140 Leningrad

Preface

There exists a tendency to divide the history of general relativity into three epochs. The first epoch started with Einstein's classical papers on the foundation of "General Relativity" (end of 1915), a field which soon opened up a vast playground for mathematicians and mathematical physicists. Though many formal aspects of Einstein's theory of gravity had successfully been tackled in this early phase, apart from various cosmological observations it was based on a few "classical tests": one test of moderate precision (Mercury's perihelion shift, approximately 1%), one test of low precision (the deflection of optical starlight, approximately 50%), and one test, with no conclusive results, that was not really a test of Einstein's field equations anyway (gravitational redshift) (Will 1984). A large number of alternative theories of gravity laid claim to viability. This situation has changed drastically in the second phase of 'testing', after about 1960. New technological innovations and techniques (atomic clocks, laser reflectors that had been placed on the lunar surface, radio interferometry, microwave techniques, etc.) not only allowed precise testing of the foundations of any physically reasonable relativistic theory of gravity (equivalence principle, gravitational redshift, etc.) and precise solar system tests of Einstein's theory, but also led to the rapid development of a new branch in physics: *relativistic astrophysics* dealing with phantastic objects such as quasars, pulsars, black holes, gravitational lenses or even the birth of our entire universe about twenty billion years ago. The discovery of the binary pulsar PSR 1913+16 by Hulse and Taylor in 1974 revealed a new arena where theories of gravity can be tested. Whereas about ten different theories of gravity so far are compatible with high-precision tests in the solar system the domain of viable theories has narrowed considerably with the radio pulse data from the binary pulsar. So far Einstein's theory of gravity has passed every experimental test with flying colors. In addition, it is obviously the simplest of all metric theories of gravity. This has finally led to the third epoch, where Einstein's theory is considered as an integral part of classical physics with its usual procedure of solving technologically oriented problems on the basis of the theoretical framework.

This monograph was written in the spirit of providing the community of people involved in the establishment, realization or use of highly precise spatial-temporal reference frames (terrestrial or extraterrestrial) or interested in the dynamics of gravitationally interacting bodies, with a useful background of general relativity. For gravitational problems in the solar system the (first) post-Newtonian approximation of Einstein's theory will usually be sufficiently accurate. We have kept the parameters β and γ from the parametrized post-Newtonian formalism *not* because of some interest in alternative theories of gravity; various combinations of these parameters merely facilitate the study of the various aspects of Einstein's theory on the post-Newtonian level. Only in Section 4.7, where the motion of extended bodies in Einstein's theory of gravity is discussed, do we depart from the post-Newtonian framework.

This monograph is divided into five Chapters. In the first Chapter the subject of 'Relativity in Astrometry, Celestial Mechanics and Geodesy' is introduced in a non-technical manner. It contains an overview of the rapid increase in measuring accuracy in these disciplines and demonstrates how various 'relativistic effects' play a role. The second Chapter presents a confrontation of Newtonian with a relativistic theory of gravity and thereby serves as a short introduction to Einstein's theory of gravity. The third Chapter deals with localization of events in time and space. Here, various practical time-scales that are already in use for diverse purposes (including, e.g. spacecraft navigation in the solar system) are discussed. Though there are many possibilities for achieving orientation in space (roads, maps etc.) in everyday life, this Chapter deals with orientation by means of gravitational fields, inertial systems such as gyroscopes and light rays originating from distant stars or extragalactic radio sources. Though localization in time and space is usually understood in an operative sense, it is often useful to relate the orientation procedure with selected coordinate systems that mainly serve for computational purposes. This problem (reference frames versus coordinate systems) is treated at the end of the third Chapter. The next Chapter is devoted to relativistic problems in celestial mechanics. It first deals with the motion of 'point-masses' such as artificial satellites about the Earth or the planets revolving about the Sun. It contains a discussion of timing observations of pulsars in binary systems and concludes with a review of the present status of the problem of motion of extended bodies in 'general relativity'. Various highly precise geodetic techniques such as superconducting gravimetry, synchronization and comparison of clocks or radio interferometry using long baselines (VLBI) are treated in Chapter five.

It is a great pleasure to thank all those people who have contributed in some way or other to the completion of this work. In particular, I would like to mention Hanns Ruder and Manfred Schneider, who have patiently followed all my trials and errors to grasp the intricacies of the present subject for more than half a decade. Thibault Damour, Jürgen Ehlers and Gerhard Schäfer are thanked for reading the whole manuscript and providing useful information, especially on the relativistic problem of motion. Special thanks go to the group of students that have worked with me here in Tübingen, notably Joachim Schastok and Eberhard Gill. Bernd Finkbeiner helped me with the typesetting system (TEX). Gratefully acknowledged is the financial support from the Deutsche Forschungsgemeinschaft (DFG); without it, this work would never have been completed.

Finally I should mention that this presentation owes much to the great works by V. A. Brumberg (1972) and C. Will (1981). In Brumberg's monograph the problem of motion is treated with great detail and care. This not only applies to the point-mass problems (lunar problem, problem of the solar system) but also to the motion of extended bodies with intrinsic angular momentum based on the hydrodynamical post-Newtonian standpoint. Unfortunately, this masterpiece has never been translated from the Russian language. Will's standard work, reflecting the spirit of the second historical phase of general relativity, provides a good theoretical background for testing theories of gravity. In the long run, however, the use of a large number of parametrized post-Newtonian parameters does not seem to be justified.

Tübingen
August 1988

Michael H. Soffel

Notation

Latin indices i, j, k, l, and so on, generally run over three spatial coordinate labels, usually, 1,2,3 or x, y, z.

Greek indices $\alpha, \beta, \gamma, \mu$, and so on, generally run over the four space-time coordinate labels $0, 1, 2, 3$ or t, x, y, z.

Repeated indices are summed unless otherwise indicated.

The signature of the space-time metric $g_{\mu\nu}$ reads: $(- + + +)$.

The components of the Riemann tensor are defined by $R^\mu{}_{\nu\lambda\sigma} = \Gamma^\mu_{\sigma\nu,\lambda} - \Gamma^\mu_{\lambda\nu,\sigma} + \Gamma^\mu_{\lambda\kappa}\Gamma^\kappa_{\sigma\nu} - \Gamma^\mu_{\sigma\kappa}\Gamma^\kappa_{\lambda\nu}$.

Einstein's equation is written as $R_{\mu\nu} - \frac{1}{2}g_{\mu\nu}R = \kappa\,T_{\mu\nu}$ with $\kappa = 8\pi G/c^4$.

Cartesian three-vectors are indicated by unslanted boldface type, like $\mathbf{a}, \mathbf{b}, \mathbf{c}$, and so on.

Four-vectors are indicated by slanted, italic boldface type, like $\boldsymbol{a}, \boldsymbol{b}, \boldsymbol{c}$, and so on.

Partial derivatives are usually denoted by commas, covariant derivatives by semi-colons. The covariant derivative w.r.t. $\boldsymbol{u}$ is denoted by $D_{\boldsymbol{u}}$.

Acknowledgment

Fig. 1.1. (p. 26) From "Tychonis Brahe Dani, Opera Omnia", I.L.E. Dreyer, Tomus V, Hauniae MCMXXIII in Libraria Gyldendaliana

Fig. 1.2. (p. 26) Left: Observatory Hoher List of the University of Bonn; right: Photograph of the Calsberg Automatic Meridian Circle, La Palma, taken by D. Calvert, Royal Greenwich Observatory

Fig. 1.5. (p. 29) HIPPARCOS; Courtesy ESA

Fig. 1.6. (p. 29) The Space Telescope; Courtesy NASA

Fig. 1.8. (p. 30) Left: LAGEOS; Courtesy NASA

Fig. 1.9. (p. 30) The GPS system; Courtesy NASA

Fig. 1.12. (p. 31) Courtesy NASA

Fig. 1.15. (p. 24) After J. Abeler, "Ullstein Uhrenbuch", Ullstein, Berlin, Frankfurt, 1975

Fig. 4.3. (p. 114) After C. Will, "Theory and Experiment in Gravitational Physics", Cambridge University Press, Cambridge, 1981

The copyright in the pictures of the LAGEOS satellite (Fig. 1.8) and the Mt. Haleakala, Hawaii, Laser Ranging Station (Fig. 1.12) does not include the jurisdictional territory of the United States of America and that the United States Government has a license for governmental purposes under any copyright.

Contents

1. Relativity in Astrometry, Celestial Mechanics and Geodesy

1.1 Astrometry

"Astrometry is the part of Astronomy that is devoted to the measurement of the positions, motions, distances, dimensions, and geometry of celestial bodies. Until the advent of Astrophysics a century ago, Astronomy consisted only of what is now called Astrometry and its theoretical counterpart — Celestial Mechanics. Practically all that was known about the Universe at the turn of the present century was obtained uniquely by astrometric techniques.

Since then, various aspects of Astrophysics have developed so tremendously, that Astrometry appeared as a rather unrewarding field of Astronomy. However, some of the very fundamental bases of our knowledge of the Universe are obtained by astrometric measurements and there is no other way to determine them. This is the case, for instance, for stellar masses, the distance scale and the dimensions of stars or galaxies. This is also the case for the kinematics of our Galaxy and, to a large extent, of its dynamics. Even if there are many spectroscopic or photometric methods to extend the distance scale to the most remote objects of the Universe, the actual scaling is based strictly on the precise measurement of trigonometric parallaxes of nearby stars." (Kovalevsky 1984).

Prior to the invention of the telescope the accuracy of astrometric measurements using instruments like a wall-quadrant or an azimuthal semi-circle (Fig. 1.1) was determined by the ability of the human eye to resolve angles and directions ($1' - 2'$). The decisive progress for the determination of astronomical angles came with the telescope.

In the year 1609 Galileo Galilei built his first telescope and saw that the Milky Way consisted of an "innumerable" number of stars, that the Moon was bespangled with mountains and craters and that Venus shows phases just

like our Moon. He discovered the sunspots and the moons of Jupiter that are
named after him. The attempt to determine the apparent diameter of Vega by
mounting a fine thread in the vertical direction and measuring the distance he
had to stand in front of the thread such that it just covered the image of Vega,
gave about 5″. Today, we know that the apparent diameter of Vega is about
1500 times smaller ($\sim$ 0.″003) and Galilei had probably measured the size of
Vega's image as produced by atmospheric scintillations.

Measuring the diameters of stars and the geometrical configuration of mul-
tiple star systems is the domain of 'very narrow field astrometry' (Kovalevsky
1984), involving angular extensions of $\lesssim$ 10″. Here, long focus telescopes or
refractors yield accuracies of the order of 0.″1. Photography of systems with
angular size $\gtrsim$ 2″ gives slightly better accuracies (0.″05 − 0.″07). The main
limitation of these visual or photographic observations results from the atmosphere. Atmospheric turbulence leads to apparent star images of minimally $\sim$ 1″
angular extension independent of the telescope's resolving power. Since atmospheric turbulent cells have a typical length scale of order 10 cm, the optical
stability time (atmospheric coherence time) is of the order of a few hundredths
of a second for wind speeds of $\sim$ 10 m s^{-1}. Using fast interferometric techniques such as speckle interferometry (Worden et al. 1976, Labeyrie 1981) the
full resolving power of large instruments can be used; speckle interferometry
frequently achieves errors of $\sim$ 0.″005 − 0.″002. Michelson interferometry with
two or more telescopes as proposed and constructed by Labeyrie (1975, 1978)
seems to be an even more powerful tool; accuracies of a few milli-arc sec seem
to be feasible.

In the angular regime of $\lesssim$ 0.5° (narrow field astrometry), where typically
star positions are linked to a certain number of surrounding stars e.g. for parallax determinations, good single observations using long-focus photography
have precisions of $\sim$ 0.″03; for certain stars systematic studies might achieve
accuracies at the milli-arc sec level.

Astrometry of star fields with relative angular separations $\lesssim$ 5° (wide field
astrometry) typically gives errors of $\sim$ 0.″2 for classical astrograph observations
and $\sim$ 0.″1 for modern Schmidt cameras (Fig. 1.2a).

Astrometry of larger parts of the sky (semi-global or global astrometry)
traditionally employs astrolabe measurements (accuracy $\sim$ 0.″1) and meridian
circles (Fig. 1.2b) with accuracies of the order of 0.″2 in each coordinate for a
single star transit. Photoelectric techniques and automatization improved this
number up to $\sim$ 0.″07.

Semi-global or global astrometry is of fundamental significance for many
branches of physics. One of the major goals of global astrometry is the mate-
rialization of a global, quasi-inertial reference frame, e.g. in form of a *Fundamental Catalogue* of stellar positions and proper motions for a larger number
of stars (or extragalactic radio sources, quasars etc.) w.r.t. some 'universal,
nonrotating, quasi-inertial coordinate system'. Generally the determination of
such nonrotating, quasi-inertial coordinates requires detailed knowledge of the

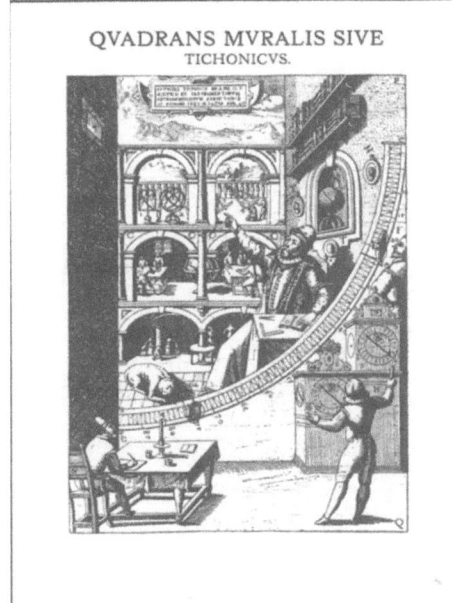
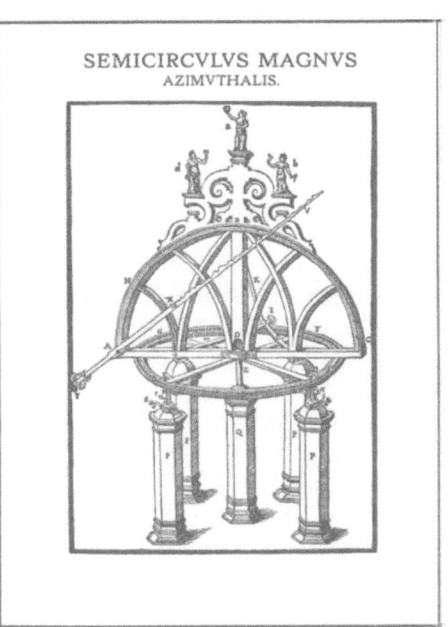

Fig. 1.1. Two Middle Aged astrometric measuring instruments. Left: wall-quadrant used by Tycho Brahe; right: great azimuthal semicircle

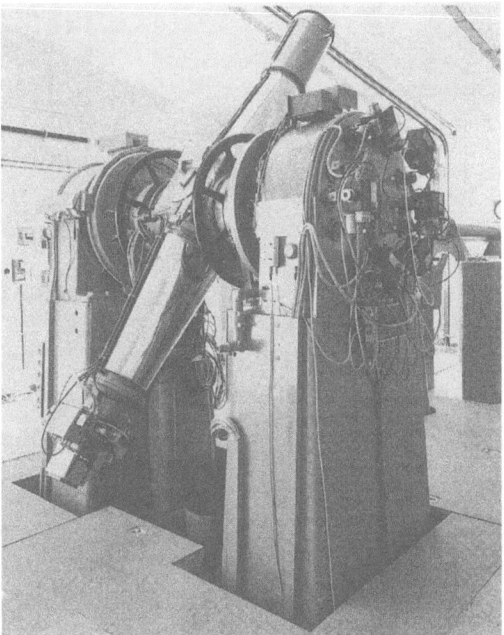

Fig. 1.2. Two modern astrometric instruments. Left: Schmidt telescope (Observatory Hoher List of the University of Bonn); right: Meridian Circle, La Palma

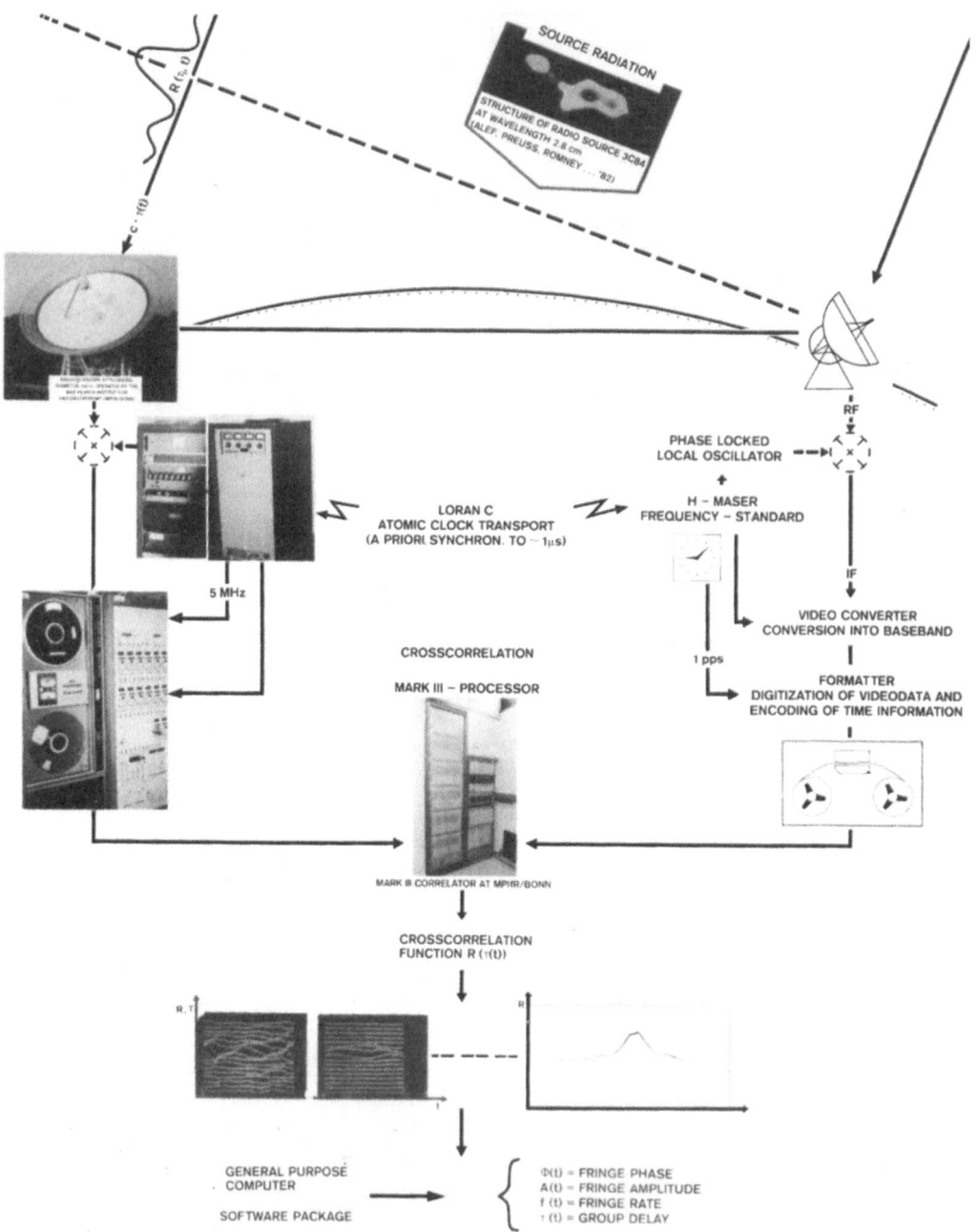

Fig. 1.3. Functional diagram of a VLBI system (Mark III)

motion of the planetary system and the rotational behaviour of the Earth in space. For example, the old astronomical quasi-inertial system documented in the FK4-catalogue was anchored to the motion of the minor planets Hebe, Iris, Flora, Metis and Eunomia. In the FK5-catalogue, where the results from solar system dynamics are corrected by a model of galactic rotation, the 'absolute rotation' is believed to be less than $\sim 1.''55$ per century (Fricke 1977, 1981).

Major progress in the field of narrow to global astrometry has been achieved by radio-interferometry using long baselines ('Very Long Baseline Interferometry' = VLBI). This technique was made possible by the development of highly stable frequency generators (atomic clocks) with frequency stabilities of better than 10^{-13} over hours and the possibility of synchronizing them to ~ 10 nsec or better. Whereas the first atomic clock, based on an absorption line of the ammonia molecule, had a stability of $\delta f / f \sim 10^{-7}$, today, frequency stabilities of $\sim 10^{-15}$ are achieved with hydrogen masers or superconducting cavity stabilized oscillators (SCSO clocks). Figure 1.3 shows a functional diagram of a two-element VLBI system, where the distance of the two observing sites might be several thousand kilometers. The radio waves, e.g. originating from an extragalactic radio source, produce a signal at the two radio antennas (Fig. 1.4) that is analyzed with the receivers typical in the GHz region. Since the two observing sites have no direct broadband connection independent oscillators (typically H-masers) are used to serve both as phase reference and for the production of time signals. The signal is then transformed into the MHz region where it is recorded on video tape together with time tags from the oscillator. The interference signal is then produced by cross-correlation of the tape data in a correlator. For astrometric and geodetic purposes the deduced quantity is the 'group delay' τ, essentially the time that has elapsed between the arrival of a certain wavefront at the two antennas. To lowest order $\tau \simeq -\mathbf{b} \cdot \hat{n}/c$, where $\mathbf{b}$ is the 'baseline vector' connecting the two antennas and $\hat{n}$ is the unit vector in the direction of the signal propagation. For the Mark III correlator system with a bandwidth of 56 MHz the total correlation accuracy is of order 0.1 nsec, corresponding to a light path of ~ 3 cm and an angular resolution of $\sim 0.''001$ for baselines of $\sim 1\,000$ km in the cm wavelength region.

As already mentioned, the traditional Earth based astrometry suffers from the disturbing influence of the Earth's atmosphere, both from atmospheric turbulence and from refraction that is difficult to model precisely, and is intimately related with the complicated rotational motion of the Earth in space. These problems are overcome with *space astrometry*.

The space astrometric mission HIPPARCOS† (Fig. 1.5) aims at the measurement of positions, parallaxes and proper motions of about 100 000 stars

---

† HIPPARCOS refers to the Greek astronomer (190 – 120 BC) who measured the position of the Moon against the stars, determined the Moon's parallax and derived the correct distance of 30 times the diameter of the Earth. He made the first accurate star map which led to the important discovery of the 'precession of the equinoxes'. HIPPARCOS is also the acronym for 'HIgh Precision PARallax COllecting Satellite'.

down to ninth magnitude with accuracies of 0.″002 or 0.″002/y. The increase in precision gained from HIPPARCOS will allow the absolute determinations, by trigonometric parallax, of distances 10 − 20 times greater than at present. Since the hierarchy of cosmic distance scales presently depends on distance measurements to the Hyades cluster, which will be improved by HIPPARCOS, it will increase our basic knowledge of the large scale structure of the universe.

The SPACE TELESCOPE (Fig. 1.6) is NASA's largest space science venture since the Viking mission to Mars. In addition to a 2.4 meter optical telescope it contains: a High Speed Photometer, a High Resolution Spectrograph, a Faint Object Spectrograph, a Faint Object Camera and a Wide Field CCD Imaging System. Three Fine Guidance sensors provide the requisite tracking of stars as faint as magnitude 14.5; one of these will be dedicated to astrometric measurements to magnitude 17. The accuracy of ST in its astrometric mode is ∼ 0.″002.

HIPPARCOS and the SPACE TELESCOPE are only the beginning of space-astrometry. The technological progress in that field is amazing indeed so that instruments with accuracies of order $10^{-4} - 10^{-6}$ arc sec are near at hand. The long focus telescope in space with 16.5 m focal length and 1 m aperture proposed by Gatewood (LMSC 1982) is expected to reach the 10^{-6} arc sec accuracy in relative position for 10 hours observing time. The same accuracy is envisaged for the Precision Optical INTerferometer in Space (POINTS) as suggested by Reasenberg et al. (1982).

An older study of 135 selected objects contained in the parallax catalogue from Yale observatory containing values for annual parallaxes derived both from trigonometrical and astrophysical determinations yielded a mean discrepancy of 0.″004 (Arifov et al. 1968) between the two determinations. This is just the value expected from Einstein's theory of gravity due to the light deflection in the gravitational field of the Sun: at the limb of the Sun this deflection amounts to $\delta\varphi \simeq 1.″75$ and decreases like $1/r$ with increasing distance r. Hence, for photon impact parameters of ∼ 1 A.U. this effect still yields

$$\delta\varphi \simeq 1.″75\, R_\odot/\text{A.U.} \simeq 1.″75 \cdot 5 \times 10^{-3} \simeq 0.″008 \; .$$

Thus, for light rays incident at about 90° from the Sun the angle of light deflection still amounts to 4×10^{-3} arc sec. Hence, for high precision astrometric measurements the light deflection in the spherical field of the Sun has to be taken into account at least at the (first) post-Newtonian level. In the parametrized post-Newtonian (PPN) formalism, the angle of light deflection is proportional to $(\gamma + 1)/2$; thus, astrometric measurements might be used for a precise determination of the *space curvature parameter* γ, whose numerical value is unity in Einstein's theory of gravity.

Historically, the light deflection in the gravitational field of the Sun had been first detected by the British expeditions to Sobral (Brazil) and Principe (Gulf of Guinea) taking photographic pictures of the solar vicinity during the solar eclipse on the 29th May, 1919. The analysis of the data gave a light

Fig. 1.4. The fundamental station Wettzell radiotelescope

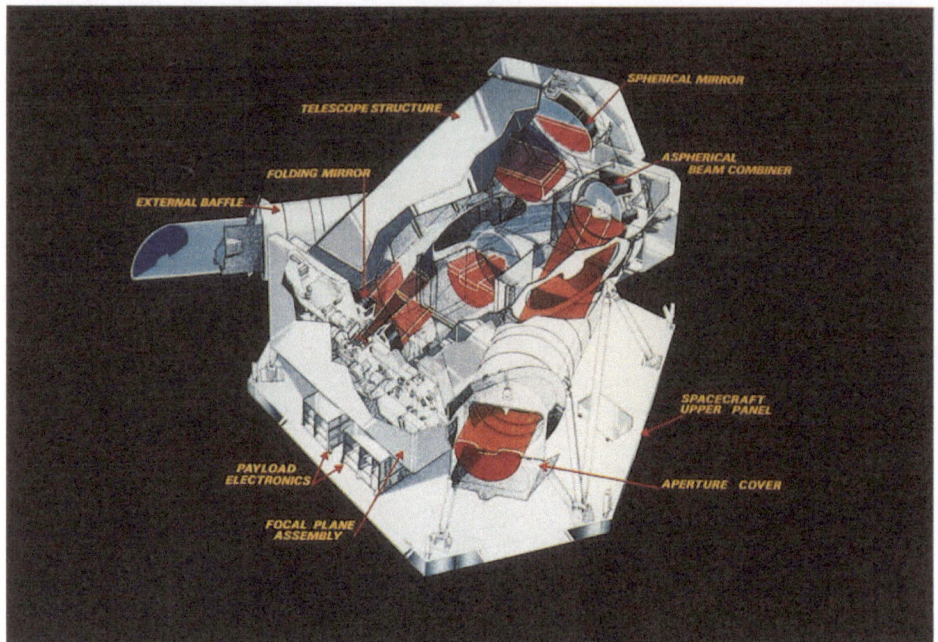

Fig. 1.5. An important future mission in space astrometry: HIPPARCOS

Fig. 1.6. The Space Telescope

deflection angle extrapolated to the limb of the Sun of 1."98 ± 0."16 (Sobral) and 1."61 ± 0."40 (Principe) in agreement with Einstein's theory (Dyson et al. 1920a, 1920b). Until now, a total of nine solar eclipses has been used for such kind of light deflection measurements. A survey over these optical measurements is presented in Table 1.1.

Table 1.1 Measurements of the Deflection of Light by the Sun.

Eclipse	Site	Number of Stars	$r/R_\odot$	$\delta\varphi$ (sec)	Ref.
May 29, 1919	Sobral	7	2–6	1.98 ± 0.16	a
	Principe	5	2–6	1.61 ± 0.40	a
Sept. 21, 1922	Australia	11–14	2–10	1.77 ± 0.40	b
	Australia	18	2–10	1.42 to 2.16	c
	Australia	62–85	2.1–14.5	1.72 ± 0.15	d
	Australia	145	2.1–42	1.82 ± 0.20	e
May 9, 1929	Sumatra	17–18	1.5–7.5	2.24 ± 0.10	f
June 19, 1936	U.S.S.R.	16–19	2–7.2	2.73 ± 0.31	g
	Japan	8	4–7	1.28 to 2.13	h
May 20, 1947	Brazil	51	3.3–10.2	2.01 ± 0.27	i
Feb. 25, 1952	Sudan	9–11	2.1–8.6	1.70 ± 0.10	j
Oct. 2, 1959	Kidal	11	1.8–8.3	2.17 ± 0.34	k
Feb. 15, 1961	Ancona	12	1.7–8.8	1.98 ± 0.46	l
June 30, 1973	Mauritania	150	1.9–9.0	1.66 ± 0.19	m

a: Dyson et al. 1920, b: Dodwell et al. 1924, c: Chant et al. 1924, d: Campbell et al. 1923, e: Campbell et al. 1928, f: Freundlich et al. 1931, g: Mikhailov 1940, h: Matukuma et al. 1940, i: van Biesbroeck 1949, j: van Biesbroeck 1953, k: Schmeidler 1962, l: Schmeidler 1984, m: Texas Mauritanian Eclipse Team 1976, Jones 1976

One sees that the results are compatible with Einstein's theory; the experimental errors, however, generally exceed the value of 10%. A number of reasons can be given for the poor quality of such ground based optical light deflection measurements. In these measurements the light deflection is determined by a comparison of the stellar images close to the Sun on the eclipse plate with the corresponding pictures of the same stellar field taken during the night (the night plate), i.e. a few months before or after the eclipse, often on the backside of the eclipse plate. The difficulty of the comparison lies in the problem of keeping the apparative conditions for eclipse- and night-plate the same to monitor and minimize possible scale changes on the photographic plate, e.g. caused by changes of the focus (see e.g. von Klüber 1960). The various difficulties involved in such optical measurements were demonstrated by the last expedition of the University of Texas (Texas Mauritanian Eclipse Team 1976; Jones 1976). The eclipse plates were taken on the 30th June, 1973 and though more than one thousand stellar images up to magnitude 10.5 should have been measurable, poor seeing reduced the number of useful stellar images drastically to 150 in the eclipse field and 60 in the reference field. Though all plots were endowed with a rectangular reference grid of artificial stellar images to monitor possible scale changes between eclipse- and comparison-field exposures the final value for the extrapolated light deflection angle was 1."66 ± 0."19 ($\gamma = 0.95 \pm 0.11$).

The situation of optical ground based light deflection measurements certainly will not be improved drastically; for space astrometry, however, the "relativity corrections" are indispensable.

In Einstein's theory of gravity the light deflection at the limb of the Sun, due to the post-post-Newtonian contributions of the spherical solar field, due to the oblateness of the Sun and due to the solar angular momentum, amount to 11×10^{-6}", $\lesssim 0.2 \times 10^{-6}$" and 0.7×10^{-6}", respectively (Epstein et al. 1980), so future long focal instruments or interferometers in space might even be in a position to detect second order gravitational bending effects.

The light deflection in the gravitational field of the Sun can be determined by means of radio interferometric techniques. Here, the spatial deflection is determined via the associated effect in the time domain, called the "gravitational time delay". The gravitational time delay is intimately related to the spatial deflection of light, since any mechanism that bends light should also "slow it down". It is really a mystery why Einstein or his contemporaries did not think of this effect; it was first treated in I. Shapiro's (1964) classical paper on a "Fourth Test of General Relativity".

The first successful radio measurements to determine the gravitational light deflection were carried out during the occultation of the radio source 3C279 in October 1969 (Seielstad et al. 1970, Muhleman et al. 1970).

Table 1.2 Older Radio Measurements of the Light Deflection by the Sun.

Observation Period	Baseline (km)	Radio Source	$\delta\varphi$ in terms of GR	Ref.
(69)9/30 to 10/15	1.06	3C279 - 3C273	1.01 ± 0.11	a
(69)10/2 to 10/10	21.57	3C279 - 3C273	1.04 ± 0.10	b
(70)10/2 to 10/15	3 900	3C279 - 3C273	1.03 ± 0.11	c
(70)10/2 to 10/26	2.7	3C279 - 3C273	0.90 ± 0.05	d
(70)9/30 to 10/15	1.4	3C279 - 3C273	1.07 ± 0.17	e
(71)10/2 to 10/15	2.7	3C279 - 3C273	0.97 ± 0.08	f
(72)10/2 to 10/13	5.0	3C279 - 3C48	1.04 ± 0.08	g
(72)10/5 to 10/12	1.4	3C279 - 3C273	0.96 ± 0.05	h
(72)9/23 to 10/20	8.45	3C279 - 3C273	0.99 ± 0.03	i
(73)10/3 to 10/12	1.4	3C279 - 3C273	1.038 ± 0.033	j
(74)3/29 to 4/25	35.3	0111-0116-0119	1.015 ± 0.011	k
(75)3/26 to 4/27	35.3	0111-0116-0119	0.999 ± 0.011	l

a: Seielstad et al. 1970, b: Muhleman et al. 1970, c: Shapiro in Weinberg 1972, d: Sramek 1971, e: Hill 1971, f: Sramek 1974, g: Riley 1973, h: Counselman et al. 1974, i: Weiler et al. 1975, j: Fomalont et al. 1975, k: Fomalont et al. 1976

In these early radio measurements the relative distance of about 9° between 3C279 and 3C273 was monitored during the time of occultation of 3C279. In this 1-frequency, 2-source technique the refractive part of the solar corona had to be modelled by means of a suitable electronic density distribution depending upon the epoch in the solar cycle (van de Hulst 1950, Blackwell et al. 1966, Blackwell et al. 1967). Since the plasma frequency ν_p in the solar corona even for $r \sim 3R_\odot$ does not exceed the value of 4 MHz, for radio measurements in the S- (2 – 4 GHz) and X-band (8 – 12.5 GHz) $\nu/\nu_p \ll 1$ and the refractive index in the corona is inversely proportional to the frequency squared ($n \simeq 1 - 0.5(\nu_p/\nu)^2$). Therefore, the frequency dependent refraction in the corona can be eliminated in a model independent manner by means of a two-frequency technique that was used since 1970. Table 1.2 (Fomalont et al. 1977), provides an overview over the older (i.e. prior to $\sim$ 1980) successful radio-light deflection measurements. In the NRAO[†] 1974/75 experiment (Fomalont et al. 1977) four elements of the NRAO with a maximal baseline of 35.3 km were used to monitor the triple system 0111+02, 0119+11 and 0116+08. The three sources are almost collinear, small in angular size, as close as 10° from each other and sufficiently bright in the radio bands. The light deflection was determined by means of relative phase measurements during the occultation period of 0116+08 around the 11th April. The two 1974/1975 experiments combined gave $\gamma = 1.014 \pm 0.018$.

The situation regarding radio light deflection measurements has been somewhat improved by the various VLBI campaigns that are run regularly over long periods of time for geodetic purposes. With a total VLBI (correlation) accuracy of $\sim$ 0.1 nsec the gravitational time delay in the field of the Sun has to be taken into account even for sources with an angular distance of almost 180° from the Sun! This is nicely demonstrated in Fig. 1.7 showing the residuals for the time delay in a VLBI-experiment from May 5 – 6, 1983 between Effelsberg and Haystack *without* (a) and *with* (b) corrections for the solar gravitational time delay. In this experiment no source was closer to the Sun than about 30°.

With such an accuracy even a time delay experiment at the limb of Jupiter (gravitational time delay of $\sim$ 1.7 nsec) appears feasible.

A detailed analysis of VLBI data from the projects MERIT and POLARIS/IRIS[‡] gave a value for the space curvature parameter (Robertson et al. 1984; Carter et al. 1985)

$$\gamma = 1.000 \pm 0.003 \ ,$$

where a formal standard error is given. It appears to be possible that regular geodetic VLBI measurements might in this way provide the best test of Einstein's theory of gravity on the post-Newtonian level.

---

† NRAO: National Radio Astronomical Observatory.

‡ MERIT: Monitor Earth Rotation and Intercompare the Techniques of observation; POLARIS: Polar-Motion Analysis by Radio Interferometric Surveying; IRIS: International Radio Interferometric Surveying.

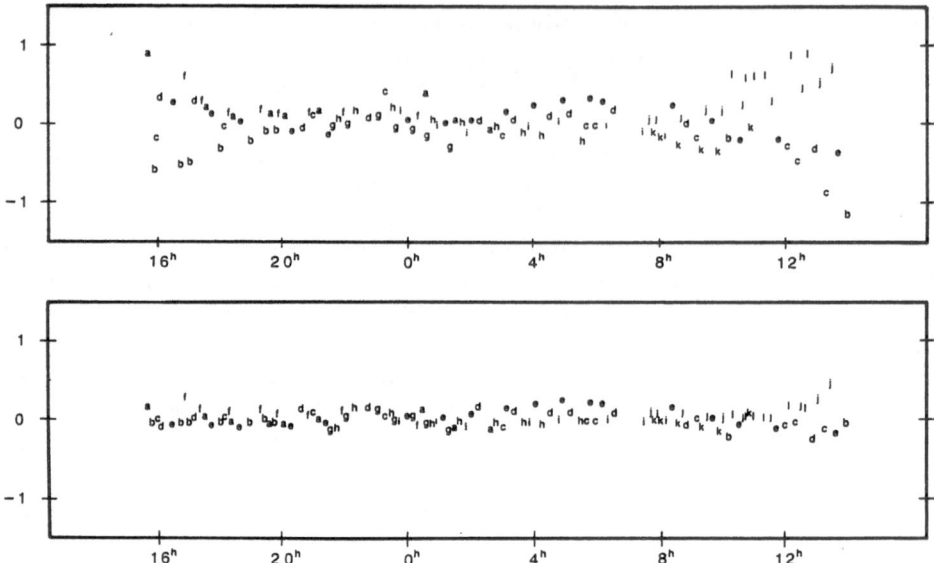

Fig. 1.7.   Residuals in nsec for $\Delta\tau$ in an Effelsberg-Haystack VLBI experiment from May 5 – 6, 1983 without (a, above) and with (b, below) corrections for the solar gravitational time delay. The 12 radio sources observed are labelled by a – l. a: 4C39.25, b:0528+134, c: 0552+398, d: 0212+735, e: 1803+784, f: 0J287, g: 3C273B, h: 0Q208, i: 3C345, j: 3C454.3, k: 2216-038, l: 0106+013

1.2 Celestial Mechanics

Whereas classically the dynamics in the solar system was determined entirely by means of astrometric observations, substantial improvement in measuring accuracy was achieved with laser and radar measurements. In satellite laser ranging (SLR) the distance to certain satellites is determined by means of laser pulses whose light travel times from the observer to the satellite and back are determined.

The most famous of these laser satellites is the LAser GEOdynamical Satellite LAGEOS (Fig. 1.8), launched into a retrograde orbit on 4th May 1976. It is a completely passive satellite of 60 cm diameter with an outer shell of aluminum and a core of beryllium copper. Its spherical surface is studded with 426 laser corner reflectors. LAGEOS's orbit is very nearly circular ($e \sim 0.004$),

Fig. 1.9. The Global Positioning System will be a constellation of 18 satellites. For navigational or geodetic purposes a total of four satellites should be in view simultaneously. Measurements of the signal travel time satellite-receiver (pseudo ranges) determine the position of the receiver with accuracies up to $\sim$10 m. Using differences between simulataneous measurements of two stations to the same satellite (single differences) several systematical errors can be eliminated leading to an accuracy of $\sim$1 m for navigation. Phase measurements can be used to determine the baseline with accuracies of a few mm

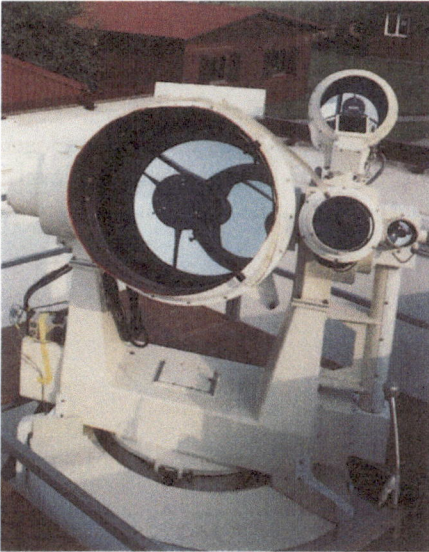

Fig. 1.8. Satellite laser ranging. Left: the geodynamical satellite LAGEOS (radius 30 cm, weight 410 kg, 426 retro-reflectors, mean height 5 900 km); right: mounting and optics of a SLR system

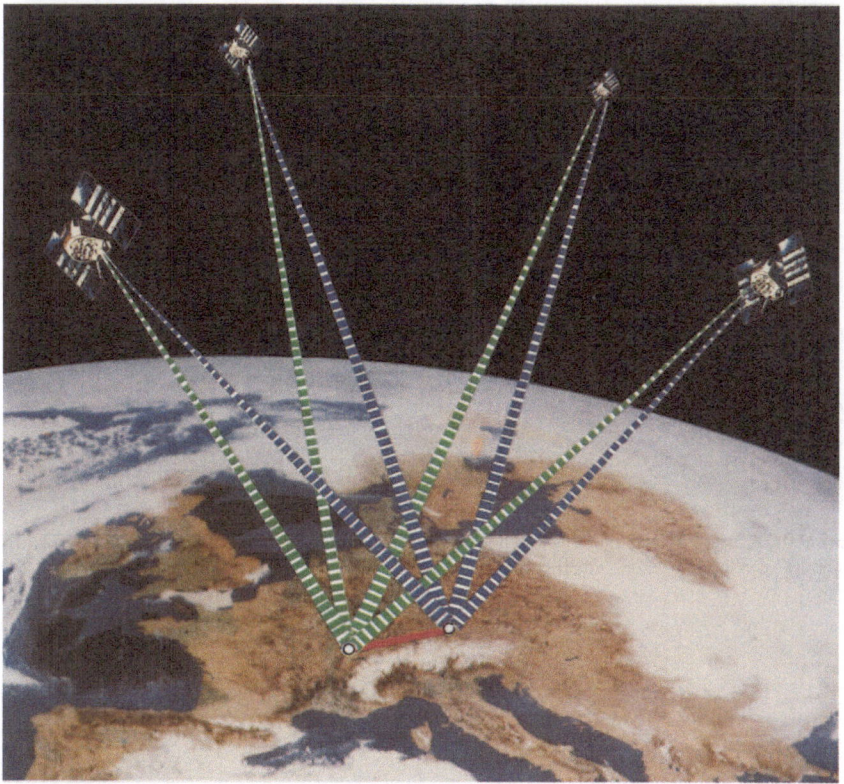

Fig. 1.9. Caption see opposite page

with a semi-major axis of about $2R_\oplus$. Other relevant data on LAGEOS can e.g. be found in Smith et al. (1980). Typically Nd-lasers with a power of $\sim 10^9$ W, working with pulse lengths of ~ 100 ps or 3 cm at a rate of a few Hz, together with an optical guidance system and computer control of the optical system are meanwhile used for satellite laser ranging. Presently, the orbit of LAGEOS can be determined at the cm level. This high precision is not only achieved by the high pulse rates and sharp temporal (1μsec), spatial ($5''$) and spectral

Fig. 1.10. A laser beam is sent to the Moon from the Mt. Haleakala station in Maui, Hawaii, as part of a lunar laser ranging experiment

(0.1 nm) filtering but also by highly advanced and reliable control of emitting and receiving telescope on the basis of precise orbital prediction codes. For the LAGEOS's orbit the measuring accuracy is mainly limited by uncertainties of atmospheric data. Since the Schwarzschild radius of the Earth ($r_s = 2GM/c^2$), representing the typical length scale for relativistic effects produced by the Earth, is about 0.88 cm relativity can no longer be ignored in the analysis of LAGEOS data. Relativistic effects in SLR affect both the motion of the satellite ("relativistic forces") and the propagation of laser pulses in the form of the gravitational time delay. Usually SLR data are analyzed in a geocentric frame. Here, the gravitational time delay in the laser propagation due to bodies other than the Earth itself (Sun etc.) can be neglected. This is no longer the case if the measurement is interpreted from a barycentric point of view. Since the computation of tidal forces is usually performed with the aid of *barycentric* ephemerides for the tide raising bodies the barycentric frame plays a role also for SLR. In the analyses of SLR data (or laser data to the Moon) geocentric and barycentric quantities have to be distinguished very carefully to avoid inconsistencies. For example in the barycentric frame the dominant relativistic effect with relative amplitude of 10^{-8} in the orbit (satellite or Moon) results from the Lorentz contraction due to the velocity of the Earth-Moon system ($v_\oplus \sim 30$ km/s) about the barycenter. This amounts to effects of the order 10 cm for the LAGEOS orbit. These effects are eliminated by means of a Lorentz transformation to a co-moving geocentric system. In the geocentric frame the relative perturbations of satellite orbits due to relativistic effects are of order r_s/r or $\sim 10^{-9}$ for low flying satellites.

In Fig. 1.11 we compare the dominant three relativistic satellite accelerations (standard post-Newtonian coordinates; geocentric frame): i) the contribution from the post-Newtonian spherical field of the Earth (dotted curve), ii) the Lense-Thirring acceleration due to the gravito-magnetic field of the rotating Earth (dashed curve) and iii) the relativistic acceleration due to the oblateness of the Earth (dotted-dashed curve) with a set of other accelerations as a function of the semi-major axis of the satellite's orbit (all other parameters as those of LAGEOS). Here the contributions from the various zonal multipole moments are indicated by the corresponding index. The tidal accelerations caused by the Moon, Sun, Venus and Jupiter are indicated by the names of the celestial bodies. For the LAGEOS orbit we also included estimates of the direct solar radiation pressure ($\square$) (e.g. Anselmo et al. 1983), the Earth's albedo ($\triangle$), infrared pressure (∇) and the charged particle ($\Diamond$) and neutral particle ($\circ$) drag (Rubincam 1982). Note that at the present level of accuracy the orbit analysis and prediction face the problem of modelling accelerations as small as $\sim 10^{-15}$ km s^{-2}.

The post-Newtonian spherical field leads to the well known Einstein perihelion precession that was first detected in the orbit of Mercury. In the post-Newtonian framework it depends upon the value for the space curvature parameter γ and a parameter β describing the amount of non-linearity contained

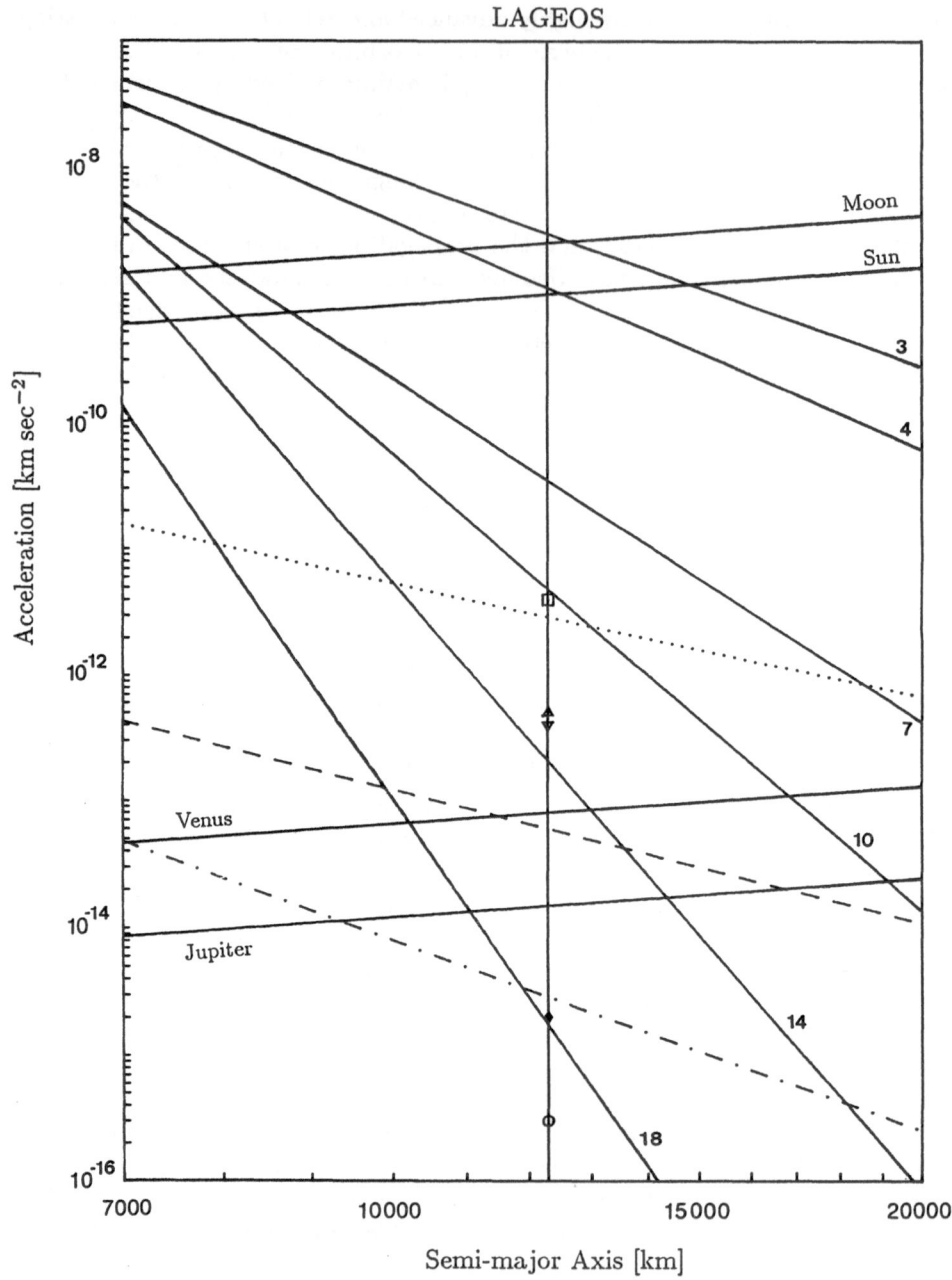

Fig. 1.11.   Various accelerations of an Earth's satellite as function of the orbit's semi-major axis. Other orbital parameters: $e \sim 0.004$, $I \sim 110°$, $\Omega \sim 197°$, $\omega \sim 72°$. The vertical line is for the semi-major axis of LAGEOS, $a \sim 12\,270$ km. The various curves are explained in the text

in Einstein's theory of gravity. Though the eccentricity of the LAGEOS orbit is extremely small and the argument of perigee badly defined, a recent analysis of LAGEOS data by Vincent (1984) suggests that Einstein's perihelion precession of the LAGEOS orbit ($\langle \Delta\omega \rangle \simeq 3''/\text{y}$) might indeed be measurable and precise values for β and γ might be inferred from LAGEOS data.

In analogy to the situation in electrodynamics matter currents (rotating masses, moving planets etc.) produce gravito-magnetic fields that in turn lead to magnetic type gravitational forces on moving matter.† The gravito-magnetic field of the rotating Earth induces an additional perihelion precession of satellite orbits and a secular drift of the nodes of the same order of magnitude. This effect was first described by Lense and Thirring (1918, Thirring 1918) and is frequently referred to as the Lense-Thirring effect. For the LAGEOS orbit this secular drift in the node is of the order of $\sim 2'' \times 10^{-5}/\text{rev.}$, roughly comparable with the effect from the $l = 12$ multipole. The discovery of the gravito-magnetic (Lense-Thirring) field of the Earth by means of the analysis of LAGEOS's orbital node presently suffers from the large uncertainties in the low mass multipole moments of the Earth. However, the Newtonian expression for the secular motion of Ω due to the multipoles to first order

$$[\Delta\Omega(Y_l^m)]_{\text{sec.}}^{(1)} = \cos I \cdot f(a, e, \text{even powers of } \sin I)$$

indicates that with a further LAGEOS satellite, placed in orbit with about the same orbital parameters as that of LAGEOS but inclination $180° - I_{\text{LAG}}$ the nodal motion from the multipoles can be eliminated by taking the sum of the two nodes and the gravito-magnetic field possibly be detected (Ciufolini 1986a, 1986b).

According to the gravito-magnetic field of the Earth the spin axis of a torque free gyroscope in orbit should precess w.r.t. fixed star oriented axes. This effect was first described by Pugh (1959) and Schiff (1960). Since the axes of torque free gyros represent local inertial axes defined by the absence of inertial forces (centrifugal and Coriolis forces) in the motion of test particles one says that the gravito-magnetic field leads to the dragging of inertial frames. For the spin axis of such a gyroscope in a low altitude polar orbit, oriented perpendicularly to the orbital plane, this precession amounts to $\sim 0.''05$ per year. Measuring that effect is one goal of the well known Stanford gyroscope experiment (NASA's gravity probe B, GPB, see e.g. Everitt 1974, Lipa et al. 1974, Worden et al. 1974, Lipa et al. 1978). In the GPB experiment (Fig. 1.12) the gyroscopes are quartz spheres with the size of a ping-pong ball, just one and a half inches in diameter. These spheres are round to an accuracy of better than a millionth of an inch.

To meet the exacting requirements of this experiment the GPB team had not only to achieve this unprecedented roundness of the spheres. They had to

---

† During the 5th Marcel Grossmann Meeting (Perth, August 1988) K. Nordtvedt (Jr.) informed me that the existence of the gravito-magnetic force can be inferred from the motion of the binary pulsar PSR 1913+16.

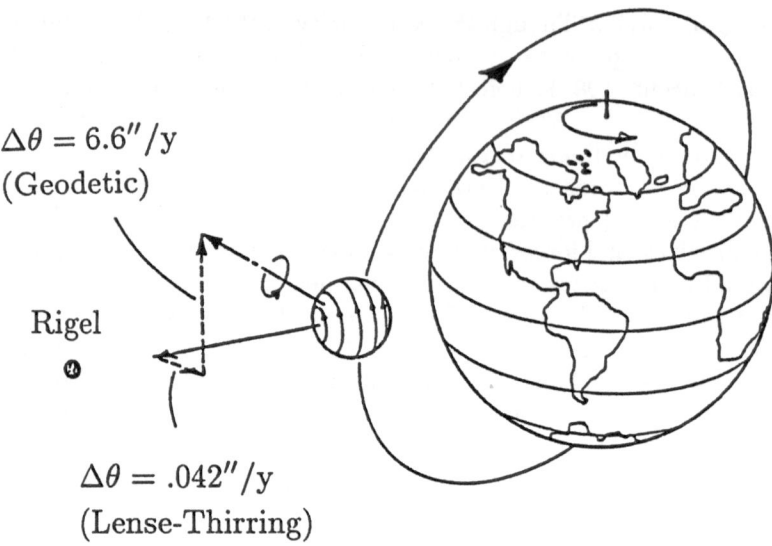

$\Delta\theta = 6.6''/y$
(Geodetic)

Rigel

$\Delta\theta = .042''/y$
(Lense-Thirring)

Fig. 1.12. The Stanford gyroscope experiment. The relativistic precession rates of gyroscope whose spin vector is parallel to the line of sight to a guide star and located in the orbital plane of a 650 km polar orbit. The geodetic precession is in the plane of the orbit and has a predicted value in Einstein's theory of 6.6 arc sec/y. The Lense-Thirring precession is in the plane of the celestial equator and in the same sense as the Earth's rotation; its predicted value is 0.042 arc sec/y

surmount other challenges that drove them to the frontier of present technology and beyond. According to the latest design a drag free satellite will house two pairs of gyros and a 4 m focal length reference telescope with folded optics, all made of fused quartz and attached to a very rigid quartz block. The telescope will point to the reference star Rigel in the constellation of Orion to provide a stable reference frame against which the gyro precession will be monitored. Interestingly enough, one of the major uncertainties of the experiments will be the proper motion of Rigel; to achieve the envisaged accuracy of about $3 \times 10^{-4''}$ (the angle subtended by the width of a human hair at a distance of 10 miles!) the optical reference direction needs the correction for the light deflection in the gravitational field of the Sun. The quartz block is cooled by liquid helium to a temperature of 1.6 degrees K. The four gyros bear coatings of superconducting niobium, are electrostatically levitated and are spun by jets of helium gas until they rotate with about 2 000 revolutions per second. After the gyros rotate with their nominal speed the helium gas is evacuated so that they will lose only a quarter of one percent of their spin rate over the period of a year. The readout of the gyros' spin axes takes advantage of the magnetic (London) moment of rotating superconductors aligned with the spin axes. A tilt of the axes can then be monitored with SQUID magnetometers.

Another relativistic effect that enters into the spin motion of the gyros in the GPB experiment is the *geodetic precession*. It is a purely kinematical post-Newtonian effect that occurs because of the gyros' motion about a central gravitating body (Earth). For a low altitude polar orbit and spin axes in the orbital plane the geodetic precession leads to a secular spin drift of $\sim 6.6''/\mathrm{y}$.

Laser ranging to retro-reflectors that had been placed on the lunar surface (Fig. 1.10) by APOLLO astronauts (APOLLO 11, 14, 15) and unmanned soviet probes (LUNA 21) with the pulse-echo method is regularly done since 1969 by the McDonald Observatory (Texas), by the station in Grasse (southern France) and the Soviet station on the Crimea. Since 1984 measurements became possible from the Haleakala station on Hawaii and since 1985 from the Australian station Orroral. This community will be joined by the German station in Wettzell in the near future. One typically works with laser-pulses of ~ 150 ps or 5 cm length; for a pulse-power of $\sim 10^8$ W this implies a number of $\sim 10^{19}$ photons per pulse for a wavelength of $\sim 5\,300$ Å. The laser beam illuminates an area of ~ 20 km^2 on the lunar surface; if it hits one of the laser reflectors ($A \sim 1$ m^2) there is a chance that one or a few photons at most return from their flight that can be detected and used for the determination of the light travel time. Important for the detection are precise frequency filtering together with the use of time gates in order to identify the relevant photons. Residuals in Lunar Laser Ranging (LLR) meanwhile tend to fall below the 10 cm level. The level of one centimeter will probably be reached in the near future. Other techniques to determine accurately the position of the Moon are radio measurements to quasars being occulted by the Moon or precise determinations of stellar occultations by the Moon.

With the rapid advance in computer technology, corresponding improvements in software and integration schemes (e.g. Fox 1984) lunar and planetary ephemerides of great internal accuracy are conveniently produced on a purely numerical basis in combination with the adjustment of initial conditions and dynamical parameters to a wealth of data. At present the numerically produced ephemeris DE200 from the Jet Propulsion Laboratory (JPL, Pasadena) is used worldwide as the standard reference. Modern analytical lunar theories also achieve great internal accuracies ($\sim 10^{-5}{}''$ in longitude and latitude and a few centimeters in distance; Henrard 1979, Chapront-Touzé 1980); however, these theories are also machine-made by the use of formal symbol manipulation and improvements in accuracy are usually connected with a tremendous increase in complexity. In modern lunar or planetary ephemerides such as DE200 the various relativistic dynamical effects are taken into account in the (first) post-Newtonian approximation, i.e. the problem of motion of the point-mass system is solved by integrating the Einstein-Infeld-Hoffmann equations of motion. Effects from the extensions, the rotations of the celestial bodies and from tidal friction are treated in a Newtonian manner. These ephemerides usually are computed in a barycentric frame. Here, the dominant oscillation in the Earth-

Moon (coordinate) distance has an amplitude of ~ 1 m and a period of half a synodic month due to the (apparent) Lorentz-contraction already mentioned above. In a suitable frame co-moving with the Earth-Moon barycenter the dominant oscillation has an amplitude of ~ 2 cm. The most important secular relativistic effect in the lunar motion is the geodetic precession. It leads to a secular drift in the lunar perigee and node of $2''$ per century as was already shown in the classical paper by de Sitter (1916). Strictly speaking, this geodetic precession rate refers to a motion w.r.t. the fixed stars; however, equally well we can refer the lunar motion to the post-Newtonian (PN) planetary dynamical frame that is determined solely by means of solar-system observations. Now, LLR data can be used to determine accurately the rate of the lunar perigee motion minus the mean motion rate for the Sun or equivalently, the mean motion of the Earth w.r.t. the planetary dynamical frame. The latter can observationally be tied to the motion of Mars and can also be determined accurately. First estimates (Bertotti et al. 1987) of the various uncertainties involved in the problem strongly indicate that the difference between observed and computed values (including relativity effects) for the rate of the lunar perigee motion is significantly smaller than the value of the geodetic precession which therefore seems to be measurable.

Data from lunar laser ranging can be used to estimate the temporal constancy of the gravitational "constant" G. Though there is a large amount of astrophysical and geophysical material (e.g. the long term behaviour of the Earth's surface temperature or length of the day etc.) that can be used for the determination of $\dot{G}/G$ LLR measurements are among the more reliable methods showing that today $\dot{G}/G$ cannot be larger than a few parts in 10^{-11} (Reasenberg et al. 1976, Anderson et al. 1978, Reasenberg et al. 1978, Hellings et al. 1983).

LLR data can furthermore be used to provide important information on the validity of the so-called "strong equivalence principle". If we write Newton's law of gravity as

$$m_I \ddot{\mathbf{r}} = -G \frac{M m_G}{r^2} \frac{\mathbf{r}}{r}$$

the *weak equivalence principle* asserts the equivalence of inertial mass m_I and gravitational mass m_G, or more precisely that the world-line of an uncharged test particle is independent of its internal structure and composition. Verification of this principle has been the concern of physicists at least since the time of Ioannis Grammaticus in the 5th Century (Shapiro et al. 1976). Laboratory experiments performed over the past 300 years with all kinds of material have shown the validity of this principle to the level of $\sim 10^{-12}$. Proposed experiments in orbit using the free fall of cylinders are expected to be sensitive at the 10^{-18} level (Worden 1978).

Einstein's Equivalence Principle (EEP) asserts the existence of local inertial frames where all *non-gravitational* laws of nature take their usual form from special relativity. For the laws of electromagnetism this e.g. implies the validity of the well known form of the first order gravitational redshift:

$$\frac{\Delta\nu}{\nu} = (1+\alpha)\frac{\Delta U}{c^2} \quad,$$

with $\alpha = 0$. Here U denotes the Newtonian gravitational potential. Mößbauer measurements over a vertical distance of $\sim 25\,\mathrm{m}$ proved this prediction to within 1%. Substantial improvement was achieved by the use of highly precise and stable atomic clocks and the possibility of time comparison over large distances using pulsed laser or microwave links with sub-nsec accuracy. In the Gravity Probe A experiment by Vessot et al. (1979, 1980, Vessot 1984) a highly stable hydrogen maser was launched to an altitude of $10\,000\,\mathrm{km}$ by a SCOUT rocket and its frequency compared to a similar clock on the ground using a microwave link. The experiment gave: $|\alpha| < 2 \times 10^{-4}$.

The strong equivalence principle finally extends the EEP to self-gravitating bodies. For example if

$$\delta_{\mathrm{N}} = (m_G/m_I)_{\leftmoon} - (m_G/m_I)_{\oplus} \equiv \eta_{\mathrm{N}}(\Delta_\oplus - \Delta_{\leftmoon}) \neq 0$$

with

$$\Delta \equiv -\frac{m_{\mathrm{G-energy}}}{m_I} \quad,$$

the Earth and Moon would fall at different rates towards the Sun leading to a relative acceleration of order

$$\delta a \sim \frac{GM_\odot}{(\mathrm{A.U.})^2}\,\delta_{\mathrm{N}}$$

and an oscillation of the Earth-Moon distance with amplitude $9.2\,\eta_{\mathrm{N}}$ m. Lunar laser ranging gave $\eta_{\mathrm{N}} = .001 \pm .015$ (Shapiro et al. 1976).

Radar measurements in the solar system, successfully performed first in September, 1959 to Venus, have significantly increased the measuring accuracies of the positions of Venus, Mercury and Mars. In a first period the propagation times of radar signals being sent towards either of the inner planets (Venus or Mercury) are measured by detecting the radar echo being reflected by the planet's surface ("passive radar"). One of the major problems of that method results from the largly unknown planetary topographies introducing errors of a few hundred meters in distance.

"The second type of target is an artificial satellite, such as *Mariners* 6 and 7, used as active retransmitters of the radar signals ("active radar"). Here topography is not an issue, and the on-board transponders permit accurate determination of the true range to the spacecraft. Unfortunately, spacecraft can suffer random perturbing accelerations from a variety of sources, including random fluctuations in the solar wind and solar radiation pressure, and random forces from on-board attitude control devices. These random accelerations can cause the trajectory of the spacecraft near superior conjunction to differ by as

much as 50 m or 0.1 μs from the predicted trajectory in an essentially unknown
way. Special methods of analyzing the ranging data ("sequential filtering") have
been devised to alleviate this problem (Anderson 1974).

The third target is the result of an attempt to combine the transponding
capabilities of spacecraft with the imperturbable motions of planets by anchor-
ing satellites to planets. Examples are the *Mariner* 9 Mars orbiter and the
Viking Mars landers and orbiters" (Will 1981).

Reducing the uncertainties introduced by the solar corona by dual frequency ranging as far as possible with this method the distance to Mars could
be determined with a precision of about six meters.

Time delay measurements from the VIKING mission have resulted in the
most precise determination of the space curvature parameter γ (Reasenberg et
al. 1979):

$$\gamma = 1.000 \pm 0.002 \ .$$

Unfortunately, the Mars VIKING lander 1 has stopped operating in November,
1982.

Interplanetary radar observations mainly made at the Haystack (Mas-
sachusetts) and Arecibo (Puerto Rico) Observatories have successfully been
used for a precise determination of Mercury's perihelion advance, which had
represented an unsolved puzzle in celestial mechanics for over half a century.
The observed value of Mercury's perihelion advance w.r.t. the moving equinox
is about 5 600″ per century. Out of this number about 5 026″/century (epoch
1900) are due to the precession of the equinoxes (i.e. the use of the astronomical
reference frame); the perihelion advances resulting from the various planetary
perturbations are indicated in Table 1.3.

Table 1.3 Perihelion Advance of Mercury.

Cause of advance	Rate (″/century)
General precession (epoch 1900)	5 025.6
Venus	277.8
Earth	90.0
Mars	2.5
Jupiter	153.6
Saturn	7.3
Others	0.2
Sum	5 557.0
Observed Advance	5 599.7
Discrepancy	42.7

There remains a discrepancy between the observed value of Mercury's
perihelion advance and the value predicted by Newton's theory of gravity by
$\sim 43″$/century. The various unsuccessful attempts to explain this number are

reviewed in the extensive but uncritical work of Roseveare (1982). This anomalous perihelion precession in the orbit of Mercury had first been discovered by Leverrier in 1845. Einstein's theory of gravity leads to a relativistic correction of the Newtonian value for Mercury's secular perihelion drift of 42.98″/century. In the PPN framework the anomalous perihelion precession is given by

$$\langle \Delta\omega \rangle = 2\pi(2\gamma - \beta + 2)\frac{GM_\odot}{c^2 a(1 - e^2)}$$

per revolution for an orbit with semi-major axis a and eccentricity e. Assuming the solar quadrupole moment to be small ($J_2^\odot \lesssim 5 \times 10^{-6}$) as indicated by measurements by Hill et al. (1975, 1986) radar measurements gave (Shapiro et al. 1976):

$$\frac{1}{3}(2\gamma - \beta + 2) = 1.003 \pm 0.005 \ .$$

Measurement of this anomalous perihelion advance have been possible also for Venus (General Relativity: 8.6″/century; observed: $8.4 \pm 4.8''$/century), the Earth (3.8 vs. 5.0 ± 1.2) and the asteroid Icarus (10.3 vs. 9.8 ± 0.8). Optical observations of Icarus can favourably be used for that task since the orbit of Icarus shows a strikingly large eccentricity ($e \simeq 0.83$) and moderately small semi-major axis ($a \simeq 1.08$ A.U.; period $\simeq 409$ days).

Modern high precision numerically produced PPN ephemerides can be used in combination with LLR data and data from radar measurements to derive precise values for β and γ and obviously also for the solar quadrupole moment. Preliminary values are (Hellings 1983):

$$\beta - 1 = (-2.9 \pm 3.1) \times 10^{-3}$$
$$\gamma - 1 = (-0.7 \pm 1.7) \times 10^{-3}$$
$$J_2^\odot = (-1.4 \pm 1.5) \times 10^{-6} \ .$$

Attempts to detect the gravito-magnetic (Lense-Thirring) field in this way so far have failed and the geodetic precession seems to be detectable only in the lunar motion.

Pulsars are believed to be rapidly rotating strongly magnetized neutron stars that emit pulsed electromagnetic radiation due to the "lighthouse" effect. For most pulsars the pulse period is remarkably constant; they are among the best clocks that we can find in nature. In the course of an extensive pulsar survey at the Arecibo Observatory in Puerto Rico in July 1974 Hulse et al. (1975) detected a new pulsar (PSR 1913+16) that showed periodic Doppler shifts in the period due to the orbital motion about an unseen companion. It was soon realized that this "single-line spectroscopic binary" provided a unique system for testing metric theories of gravity at a level even higher than post-Newtonian. One reason for this lies in the high accuracy that meanwhile can be achieved in pulsar timing. For example for the fast pulsar PSR 1937+214 with a period of 1.558 ms the minute period change of $\dot{P} = 1.2 \times 10^{-19}$ could be determined observationally. One further reason is that PSR 1913+16 is a *close* binary system: from the orbital period of $\sim$ 7.75 hours and reasonable masses for the pulsar and its companion one infers from Kepler's third law that the size of the orbit is $\sim 10^{11}$ cm $\sim R_{\odot}$. For such an orbit with $v/c \sim 10^{-3}$ and $GM/c^2 r \sim 10^{-6}$ the value of the Einstein perihelion precession amounts to ~ 4.2 deg/y or $\sim 42''$ within a day. Note that it needs a whole century until the corresponding relativistic perihelion shift in the orbit of Mercury has built up to that value!

The accuracy with which the various parameters of this binary system can be determined by means of pulse-timing is really remarkable as can be seen from Table 1.4 (data taken from Weisberg et al. 1984).

Table 1.4 Orbital Parameters of PSR 1913+16.

Projected semi-major axis	$a_p \sin I = 2.341\,85 \pm 0.000\,12$ light sec
Eccentricity	$e = 0.617\,127 \pm 0.000\,003$
Orbital period	$P_b = 27\,906.981\,63 \pm 0.000\,02$ s
Longitude of periastron	$\omega_0 = 178.864\,3 \pm 0.000\,9$ deg
Julian ephemeris date of periastron and reference time for P_b and ω_0	$T_0 = 2\,442\,321.433\,208\,4 \pm 0.000\,001\,2$
Mean rate of periastron advance	$\langle \dot{\omega} \rangle = 4.226\,3 \pm 0.000\,3$ deg yr^{-1}
Gravitational redshift and time dilation	$\tilde{\gamma} = 0.004\,38 \pm 0.000\,12$ s
Orbital period derivative	$\dot{P}_b = (-2.40 \pm 0.09) \times 10^{-12}$ s s^{-1}
Orbital inclination	$\sin I = 0.76 \pm 0.14$

Weisberg et al. (1984) classify the parameters in the upper part of Table 1.4 as "classical", those in the lower part as "relativistic". Whereas a purely Newtonian analysis of the pulse-arrival time data can determine only the "classical parameters", which have the same names and essentially the same meanings as their Newtonian counterparts, a relativistic analysis, carried out to a precision consistent with present experimental uncertainties, can determine another four parameters of the system: $\langle \dot{\omega} \rangle$, the average rate of rotation of the orbital ellipse within its plane; $\tilde{\gamma}$, the amplitude of delays caused by variations in gravitational redshift and time dilation as the pulsar traverses its elliptical orbit; $\dot{P}_b$, the time derivative of orbital period and $\sin I$, where I is the inclination between the plane of the orbit and the plane of the sky (perpendicular to the line of sight). The last parameter, though cast in terms of geometry, actually is derived from the gravitational time delay caused by propagation of the pulse signal through the gravitational field of the companion. This measurement of $\sin I$ marks the first successful observation of gravitational time delay outside the solar system

(Weisberg et al. 1984). The various theoretical curves for $\dot{P}_b, \sin I, \tilde{\gamma}$ and $\langle \dot{\omega} \rangle$ as functions of the two stellar masses lead to a precise determination of the pulsar's and companion's mass. Taylor (1987) finds $m_p = 1.445 \pm 0.007$ and $m_c = 1.384 \pm 0.007$ solar masses representing the most accurate determination of the mass of a (pulsating) neutron star.

The PSR 1913+16 system became famous because it provides the first compelling evidence for the existence of gravitational radiation. General relativity predicts that a pair of masses in mutual orbit should gradually spiral closer together as the system loses energy in the form of gravitational radiation. According to Einstein's theory of gravity one expects an orbit period derivative $\dot{P}_b = (-2.403 \pm 0.002) \times 10^{-12}$. The observations through 1984 gave $\dot{P}_b = (-2.40 \pm 0.09) \times 10^{-12}$ (Weisberg et al. 1984) in excellent agreement with the theoretical value. Most relativistic theories of gravity other than Einstein's theory strongly conflict with that result (e.g. Will 1981).

1.3 Geodesy

The history of geodesy, according to the famous Friedrich Robert Helmert (1843 – 1917) the science "of the measurements and mappings of the Earth's surface" can roughly be devided into four phases.

Phase A (Spherical Earth model ; 200 BC – middle of the 17th century)

Initially one aimed at scientifically supporting and quantifying some concept of the overall shape of the Earth. In antiquity the concept of the (approximately) spherical shape of the Earth (Pythagoras, Aristoteles) had to compete with the picture of an Earth's disk being floated around by the world's ocean Okeanos. The founder of scientific geodesy is regarded as Eratosthenes (276 – 195 BC) who experimentally determined the Earth's radius assuming a spherical model of the Earth.

"Eratosthenes found that at the time of summer solstice the sunbeams in Syene (the present Assuan) fell perpendicularly into a well while in Alexandria, lying approximately on the same meridian, they formed an angle with the direction of the plumb line that he, using a shade-stick, determined to be 7° 12'. He estimated the distance between Syene and Alexandria from the camel travel time of 50 days and a day's journey of 100 stadia to be 5000 stadia. With the length of the Attic stadium ($\simeq$ 185 m) we obtain an Earth's radius of 7360 km, a value that deviates by about +16% from the radius of a mean terrestrial globe" (Torge 1975). Later, Posidonius (135 – 51 BC) determined the Earth's radius by measuring the meridian arc between Alexandria and Rhodes with an accuracy of about 11%.

Substantial progress in the determination of the shape of the Earth was not attained until the invention of the Keplerian telescope (1611) and the method of triangulation (1589) by Tycho Brahe, first applied by Snellius. By this means Snellius already in 1615 achieved an accuracy of 3.4%. Abbé J. Picard in 1669/70 even achieved an accuracy of 0.1%; for the first time he employed for his measurements a telescope with a reticule.

Phase B (Ellipsoidal Earth model; middle of 17th century – middle of 19th century)

In 1666 J. D. Cassini for the first time observed the oblateness of Jupiter at its poles. By means of an analysis of pendulum oscillations in 1672 J. Richter discovered that the Earth's gravity increases from the equator to the poles. From a theoretical point of view Isaac Newton (1643 – 1727) and Christian Huygens (1629 – 1695) by studying the equilibrium configurations of rotating fluids came to the conclusion that the overall shape of the Earth should be a rotational ellipsoid. The oblateness of such an Earth-ellipsoid has to be considered as a characteristic target quantity in this phase.

Phase C (Geoid-model; middle of 19th century – middle of 20th century)

In the first half of the 19th century P.-S. Laplace (1802), C. F. Gauss (1828), F. W. Bessel (1837) and others discovered that in general the physical direction of the plumb line does not coincide with the computed normal of the ellipsoid, and hence the rotational ellipsoid represents only a rough "model" for the global figure of the Earth. Accordingly, the ellipsoid was replaced by the geoid as level surface of constant geo-potential (gravity and centrifugal potential) at "mean sea level", extended below the continents. The determination of the geoid was therefore the main goal of geodesy for a long period (1880 – 1950) and will remain one of the major goals of geodetic reasearch in the future.

Phase D (Dynamical and relativistic model of the "Earth system"†; since ~ middle of 20th century)

With the development of satellite geodesy, laser distance measurements to the Moon and radio-interferometry with long baselines (VLBI) a fundamental change in the concept of geodesy appears in senses that might be characterized by the three pairs of notions: i) static → dynamical, ii) Earth → Earth system and iii) "Newtonian " → "Einstein's theory of gravity".

The first two changes might be illustrated with the example of changes in the length of the day (l.o.d.). Whereas in the early years the Earth's rotation was a paragon of uniform motion and served for the definition of the second as the 86 400th part of the sidereal l.o.d., in 1934/35 A. Scheibe and U. Adelsberger by means of an electric quartz clock for the first time detected seasonal changes of the l.o.d. of some 10^{-3} s (ms) per day. Table 1.5 shows that a wealth of physical interactions leads to changes of the l.o.d. The time scale therein covers

---

† "Earth system" stands for: Earth and its planetary environment.

Table 1.5 Variations in the Length of the Day

Time Scale	Dynamical Processes
10^6 y Long period variations	tidal friction angular momentum transfer into lunar orbit solar winds displacement of large masses plate motion core-mantle interaction post-glacial mass displacements melting of ice masses sea level variations formation of mountains
100 y Decade fluctuations 5 y	tides core rotation core-mantle interaction solar activities long period mass variations in the atmosphere
2 y Annual variations 0.5 y	tides solar activities wind circulations ocean currents ground-water displacememts air-water displacements sea level variations
35 d High frequency variations	tides meteorological noise global zonal winds

such long periods during which larger Earth masses are relocated ($\gtrsim 10^6$ y), down to a few days where e.g. meteorological noise and the high frequency parts of the tides are operative.

The various causes of the l.o.d. variations are: i) exterior causes (tides, solar winds etc.), ii) inner causes (core rotation, core-mantle interaction etc.) and iii) surface processes (winds, ocean currents etc.).

Here, the tides as dominant exterior forces play a crucial role leading to the necessity to take the whole *Earth system* into account instead of just an isolated Earth.

The present high precision geodetic techniques can roughly be divided into *terrestrial* and *astronomical*, where the problem of frequency and time generation and dissemination has features of both classes.

Table 1.6 gives an overview over the various terrestrial measuring techniques, their accuracy and the magnitude of expected relativistic effects that enter directly into the observables.

As one infers from Table 1.6 at present only gravimetry is able to pene-

Table 1.6 Terrestrial Geodetic Measurements and Relativistic Effects

Kind of measurement	Method	Accuracy	Magnitude of relativistic effects
Distance measurements	rigid rods Vaisäla comparator electromagnetic electro-optical	$10^{-6}D$ $10^{-7}D$ $10^{-6}D$ $10^{-6}D$	$10^{-9}D$
Horizontal directions	theodolite	$0.3''$	—
Vertical angles	theodolite	$4''$	$3 \times 10^{-4}''(D/R_{\oplus})$
height meas.	levelling	$\sim 10^{-1}''$	$3 \times 10^{-4}''(D/R_{\oplus})$
Gravity field determinations	gravimetry free fall methods pendulum gradiometry	$10^{-10}\,g$ $10^{-9}\,g$ $2 \times 10^{-7}\,g$ $\sim 1\,\mathrm{E\,Hz}^{-1/2}$	$10^{-9}\,g$ $10^{-9}\,g$ $10^{-9}\,g$ $3 \times 10^{-6}\,\mathrm{E}$

D: distance to be measured; E: Eötvös unit ($1\mathrm{E} = 10^{-9}\mathrm{s}^{-2}$).

trate into the relativistic regime, due to the development of superconducting gravimeters with an accuracy better than $10^{-10}\,g \simeq 0.1\mu$gal, corresponding to an accuracy in height of $\Delta h \simeq 0.03(\Delta g/g) \times 10^{10}$ cm. In such an instrument usually a pair of current-carrying coils at liquid helium temperature levitates a small superconducting sphere in its magnetic field; the position of the sphere is kept constant by using a capacitance transducer in a feed-back loop (Fig. 1.13). The voltage output of the feed-back system then gives a measure of the gravity variations.

The situation is drastically different for the comparison of frequency and time where relativity plays an important role. Figure 1.14 gives an overview over the art of clock making from 3000 BC until now. As already mentioned frequency stabilities of the order of 10^{-15} can at present be achieved with hydrogen masers.

In contrast to a Cs- or Rb-frequency normal the hydrogen maser clock is an active system where a hyperfine transition of frequency ~ 1.42 GHz is used to tune accurately a quartz oscillator. The essential constitutents of a H-maser: an atomic beam source for hydrogen atoms, a deflecting magnet, storage bulb and microwave resonator are depicted in Fig. 1.15.

According to the gravitational redshift the rates of two atomic clocks at different height h above sea level differ by a factor of $\sim g\Delta h/c^2$ or $\sim (\Delta h/1\,\mathrm{km}) \times 10^{-13}$, leading to an accumulated time difference of a few μs/year for $\Delta h \sim 1$ km. It is obvious that for the realization of a "world time" as reference time scale for many (geodetic, navigation etc.) measurements one has to "correct" for gravitational redshift effects. If time comparison is done with moving clocks either by clock transport or by satellite clocks then also Doppler effects have to be taken into account. For example in the satellite

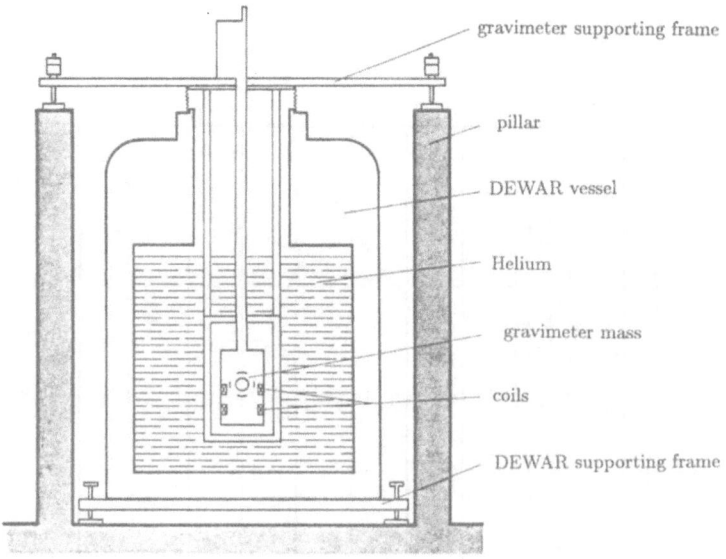

gravimeter supporting frame

pillar

DEWAR vessel

Helium

gravimeter mass

coils

DEWAR supporting frame

Fig. 1.13. Schematic diagram of a modern superconducting gravimeter

clocks of the *Global Positioning System* (GPS; Fig. 1.9) the nominal frequency $(f_0 = 10.23$ MHz) has been reduced by a factor of $(1 + 4.45 \times 10^{-10})$ prior to launch to compensate both for gravitational redshift and Doppler effects.

Modern geodetic space techniques include satellite methods such as satellite laser ranging (SLR), Doppler measurements, altimetry, satellite gradiometry or Satellite-to-Satellite Tracking (SST), lunar methods such as Lunar Laser Ranging (LLR) and astrometrical measurements like star observations and Very Long Baseline Interferometry (VLBI). Analysis of nearly ten years of LLR observations have yielded an impressive quantity of physical parameters of the Earth-Moon system with astronomical, geophysical and cosmological implications. Among the parameters determined by LLR are: Observatory coordinates, reflector coordinates, baselines, mass of the Earth-Moon system, libration angles of the Moon, lunar moment of inertia ratios, lunar gravity field coefficients, changes in the mean motion of the Moon ($\dot{n}_{\text{(}}$, caused by tidal friction), dissipation factors, lunar Love number, long period terms in the Earth rotation, universal times (UT0, UT1) and polar motion, dynamical equinox and obliquity of the ecliptic, tectonic motion and continental drifts, the Nordtvedt parameter and the value for $\dot{G}/G$. SLR, altimetry, satellite gradiometry and SST techniques are mainly used to derive accurately the global structure of the Earth's gravity field. Modern "Earth models" of the global gravity field are complete to degree and order 36 in a representation by spherical harmonics. The importance of relativity in SLR, LLR and VLBI has already been stressed above. For the possibility to use satellite gradiometry for relativity experiments in the future see e.g. Paik et al. (1988).

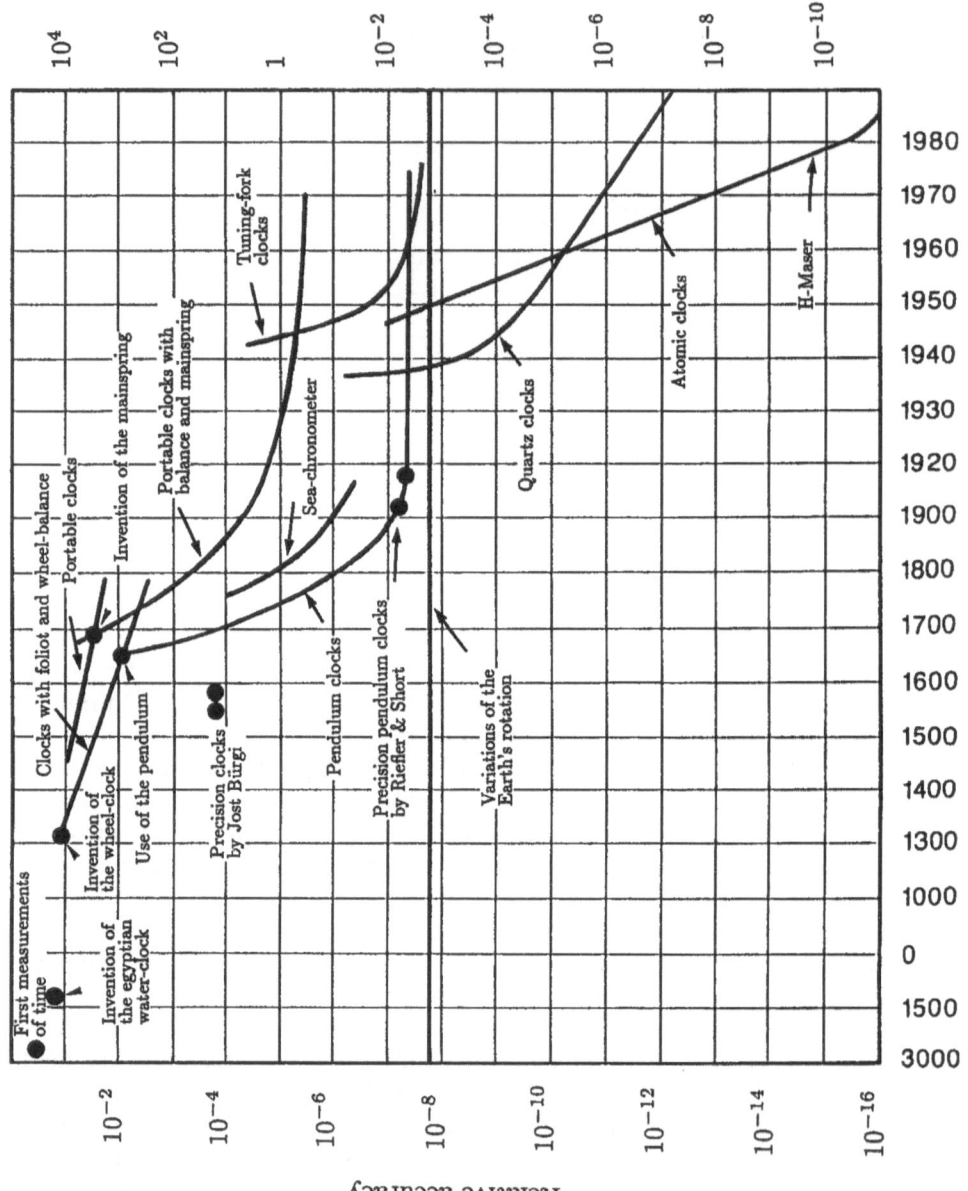

Fig. 1.14. The art of clock making from 3000 BC until now

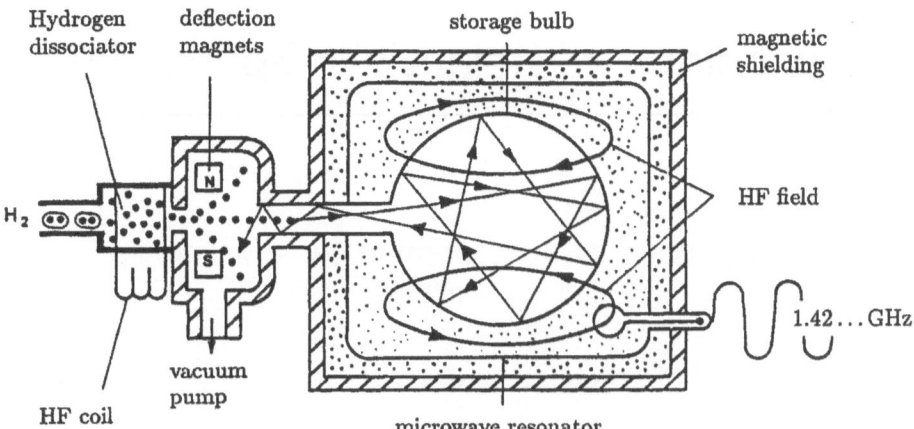

Fig. 1.15. Schematic diagram of a H-maser. Molecular hydrogen is channeled into a glass vessel where it is electrically dissociated into individual atoms. Passing through an inhomogeneous magnetic field they are separated according to their magnetic dipole moment and the atom in the higher hyperfine level are focussed into the storage bulb. The storage bulb, embedded in a microwave resonator, allows the atoms to stay in the resonance region up to about ~ 1 s without loss of energy. The resonator is radiated with a HF field of the hyperfine transition frequency to induce stimulated emission until the resonator oscillates in resonance. The resonance frequency can then be used to tune accurately a quartz oscillator

2. Newtonian and Relativistic Space-Time

2.1 Newtonian Space-Time

Let us start with a close inspection of Newton's theory of gravity (see also the representations by Cartan (1923, 1924); Havas (1964); Trautman (1965); Misner (1969); Misner et al. (1973)).

Let us consider a closed system of N mass points with a 2-body force of the form:

$$\mathbf{F}_{ij} = \mathbf{x}_{ij} \cdot f_{ij}(|\mathbf{x}_{ij}|) \quad ; \quad \mathbf{x}_{ij} = \mathbf{x}_i - \mathbf{x}_j \ , \tag{2.1.1}$$

obeying the law of "actio = reactio":

$$\mathbf{F}_{ij} = -\mathbf{F}_{ji} \ . \tag{2.1.2}$$

If $\mathbf{x}_k(t)$, $k = 1, \ldots, N$ represents a solution of the equations of motion

$$m_i \ddot{\mathbf{x}}_i = \sum_{j \neq i} \mathbf{F}_{ij} \quad (1 \leq i \leq N) \tag{2.1.3}$$

then also

$$\mathbf{x}'_k(\pm t + b) \equiv R\,\mathbf{x}_k(t) + \mathbf{v}t + \mathbf{d} \ . \tag{2.1.4}$$

Here, $\mathbf{v}$ and $\mathbf{d}$ are constant vectors, $b \in \mathbb{R}$ and R is a rotational matrix with $R^T R = 1$.

Hence, the Newtonian equations of motion (2.1.3) are invariant under *Galilean transformations* of the form:

$$\mathbf{x}' = R\,\mathbf{x} + \mathbf{v}t + \mathbf{d} \tag{2.1.5a}$$

$$t' = t + b \ , \tag{2.1.5b}$$

forming the 10-parameter (parameters: $(R, \mathbf{v}, \mathbf{d}, b)$) *Galilean group*. *Newton's absolute time t*, that appears here, might be considered as a smooth mapping of a 4-dimensional space-time manifold $\mathcal{M}$ into the real numbers. It is uniquely determined up to linear transformations (origin and unit) of the form

$$t \rightarrow at + b \quad ; \quad a \in \mathbb{R}^+ \, , \quad b \in \mathbb{R} \, .$$

This absolute time t induces a *Newtonian foliation* of $\mathcal{M}$: it cuts the manifold into absolute simultaneity- or space-sections, defined by all points $p_\alpha \in \mathcal{M}$ with

$$t(p_\alpha) = t(p_\beta) \, .$$

An ideal clock in the Newtonian sense then might indicate the duration between two events in $\mathcal{M}$, $t(p_2) - t(p_1)$. On the space-sections induced by t the Euclidean geometry with a Euclidean metric g_3 is valid.

The *Galilean space-time* characterized by these properties can be understood as a 4-dimensional vector-space; i.e. for any two points $p_1, p_2 \in \mathcal{M}$ there exists a uniquely determined difference vector $\mathbf{p_1 p_2}$ from p_1 to p_2. A local reference frame on the Galilean space-time is then given by the "world line" $\gamma : (t(\lambda), \mathbf{x}(\lambda))$ of an observer, defining the origin of the reference frame and a triad $\mathbf{e}_{(i)}$ $(i = 1, 2, 3)$ of vectors that are orthonormal w.r.t. g_3 and lie in the space-sections. They define the spatial reference directions. Each of these reference frames induces a coordinate system (see Fig. 2.1):

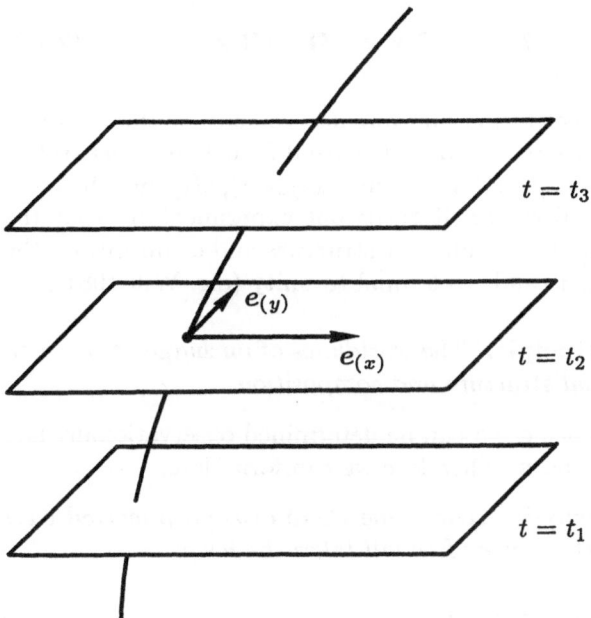

Fig. 2.1. Absolute foliation of the Galilean space-time and Newtonian reference frames

Define $t = 0$ for some point $\mathcal{O} \in \gamma$. Then let $x^0 = 0$ on the space-section through $\mathcal{O}$. Choose units of time and length once and for all. Then for an event $p \in \mathcal{M}$ let $x^0(p) = t(p)$ and $x^i(p)$ in accordance with the affine vector space structure:†

$$\mathbf{op} = x^i(p)\, \mathbf{e}_{(i)}(p') \ ,$$

where p' denotes the point in γ simultaneous to p (Fig. 2.1).

One then finds that the Galilean space-time admits a preferred class of reference frames with induced coordinates (Galilean coordinates) where Newton's law for the motion of a point particle takes the usual form:

$$\mathbf{F} = m_I \ddot{\mathbf{x}} \ . \tag{2.1.6}$$

Here the index I of m refers to the *inertial* mass of the accelerated body and the force $\mathbf{F}$ can directly be derived from the interaction with the environment. However, this definition of Galilean systems becomes problematic if gravitational fields are taken into account (e.g. Ehlers 1973). In that case (2.1.6) can be replaced by

$$m_I \ddot{\mathbf{x}} = m_G \mathbf{g} + \mathbf{F'} \ , \tag{2.1.7}$$

in a Galilean system, where m_G denotes the gravitating mass of the test body and $\mathbf{g}$ the gravitational field strength. In an arbitrary reference frame with associated coordinates this equation takes the form

$$\ddot{\mathbf{x}} = \frac{m_G}{m_I}\mathbf{g} + \frac{1}{m_I}\mathbf{F'} - \mathbf{a} - 2\,\mathbf{\Omega} \times \dot{\mathbf{x}} - \dot{\mathbf{\Omega}} \times \mathbf{x} - \mathbf{\Omega} \times (\mathbf{\Omega} \times \mathbf{x}) \ , \tag{2.1.8}$$

where $\mathbf{a}$ and $\mathbf{\Omega}$ denote the acceleration and angular velocity of the reference frame w.r.t. a Galilean one. If there were objects with different (m_G/m_I)-ratios then in a free fall experiment one could determine $\mathbf{x}(t), \dot{\mathbf{x}}(t), \ddot{\mathbf{x}}(t)$ and therefore $\mathbf{g}, \mathbf{a}, \mathbf{\Omega}$ and m_G/m_I directly. However, it turns out experimentally that the ratio m_G/m_I is identical for bodies of different structure and composition (the weak equivalence principle) and can be set equal to unity (e.g. Will 1981).

The Weak Equivalence Principle. *The worldlines of uncharged test particles are independent of internal structure and composition.*

Therefore only the difference $\mathbf{g} - \mathbf{a}$ can be determined observationally and not the separate values for $\mathbf{g}$ and $\mathbf{a}$. Therefore we can formulate:

Law of Galileo. *In the Newtonian space-time there exists a preferred class of reference frames in which the law for free fall takes the form:*

$$\ddot{\mathbf{x}} = \nabla U \ . \tag{2.1.9}$$

---

† Summation over equal indices.

Here the gravitational potential U is a function depending on $(t, \mathbf{x})$ and the reference frame.

These equations of motion can now be written as follows:

$$\frac{d^2 x^\mu}{d\lambda^2} + \Gamma^\mu_{\nu\sigma} \frac{dx^\nu}{d\lambda} \frac{dx^\sigma}{d\lambda} = 0 \quad ; \quad \frac{d^2 t}{d\lambda^2} = 0 \; , \tag{2.1.10}$$

if we set:

$$\Gamma^i_{00} = -\frac{\partial U}{\partial x^i} \equiv -U_{,i} \quad ; \quad \Gamma^\mu_{\nu\sigma} = 0 \text{ otherwise} . \tag{2.1.11}$$

Instead of saying that gravity leads to a deviation from a "straight test body trajectory", we can comprehend the worldline of a test body as a *geodesic* in a curved space-time with affine connection Γ. The Newtonian connection is symmetric ($\Gamma^\mu_{\nu\lambda} = \Gamma^\mu_{\lambda\nu}$) and leads to a non-vanishing (Newtonian) curvature tensor

$$R^\mu_{\ \nu\lambda\sigma} = \Gamma^\mu_{\sigma\nu,\lambda} - \Gamma^\mu_{\lambda\nu,\sigma} + \Gamma^\mu_{\lambda\kappa}\Gamma^\kappa_{\sigma\nu} - \Gamma^\mu_{\sigma\kappa}\Gamma^\kappa_{\lambda\nu} \tag{2.1.12}$$

with

$$R^i_{\ 0j0} = -U_{,ij} \quad ; \quad R^\mu_{\ \nu\lambda\sigma} = 0 \text{ otherwise} . \tag{2.1.13}$$

This just describes the Newtonian tidal forces. To see this we consider a whole bundle of freely falling particles whose motion is given by $x^i(t, \sigma)$. In *Newtonian coordinates* we can write:

$$\ddot{\mathbf{x}}|_\sigma - \nabla U = 0 \tag{2.1.14}$$

or

$$\frac{\partial}{\partial \sigma} \left(\frac{d^2 x^i}{dt^2} - \frac{\partial U}{\partial x^i} \right) = 0 \; . \tag{2.1.15}$$

With

$$\frac{\partial}{\partial \sigma} = \frac{\partial x^i}{\partial \sigma} \frac{\partial}{\partial x^i} \equiv n^i \frac{\partial}{\partial x^i}$$

we obtain

$$\frac{\partial^2 n^i}{\partial t^2} - \frac{\partial^2 U}{\partial x^i \partial x^j} n^j = \frac{\partial^2 n^i}{\partial t^2} + R^i_{\ 0j0}\, n^j = 0 \; , \tag{2.1.16}$$

where $n^i = \partial x^i / \partial \sigma$ might be visualized as the vector connecting "neighbouring geodesics". This equation of "geodetic deviation" just expresses the fact that tidal forces can be understood as the action of a curvature tensor. This, together with the possibility of describing the orbit trajectories of freely falling test-particles as geodesics, results from the (weak) equivalence principle.

So far we have used Galilean (Newtonian) coordinates for the formulation of the equations of motion. It is, however, not difficult to formulate the structure of Newtonian space-time geometrically, i.e. independently of coordinates: e.g. the geodetic equation (2.1.10) can be written as

$$D_u u = 0 \quad ; \quad u^\mu_{\ ;\nu} u^\nu = 0 \; , \tag{2.1.17}$$

where $u^\mu = dx^\mu/d\lambda$ denotes the "test particle's 4-velocity" and D (or the semi-colon in the coordinate representation) the covariant derivative w.r.t. Γ:

$$u^\mu{}_{;\nu} = u^\mu{}_{,\nu} + \Gamma^\mu_{\nu\sigma} u^\sigma \ . \tag{2.1.18}$$

Let $n = \partial/\partial\sigma$, then (2.1.14) reads:

$$D_n(D_u u) = 0$$

or

$$(D_u D_n + [D_n, D_u])u = 0 \tag{2.1.14'}$$

with

$$[D_n, D_u] \equiv D_n D_u - D_u D_n \ .$$

Because of the symmetry of the connection coefficients:

$$D_n u - D_u n = [n, u] = \left[\frac{\partial}{\partial\sigma}, \frac{\partial}{\partial\lambda}\right] = \frac{\partial^2}{\partial\sigma\partial\lambda} - \frac{\partial^2}{\partial\lambda\partial\sigma} = 0$$

and therefore

$$D_u D_u \, n + [D_n, D_u] \, u = 0 \ .$$

Since $[n, u] = 0 = D_{[n,u]}$ we obtain:

$$D_u D_u \, n + \mathcal{R}(n, u)u = 0 \ , \tag{2.1.15'}$$

where

$$\mathcal{R}(n, u) = [D_n, D_u] - D_{[n,u]} \tag{2.1.19}$$

denotes the mapping corresponding to the curvature tensor (e.g. Misner et al. 1973). Equation (2.1.15') expresses in a geometrical way how the (Newtonian) curvature tensor determines the tidal forces.

Equation (2.1.11) reveals that the curvature is one of *space-time* and not one of the space sections. On these we have $\Gamma^i_{jk} = 0$, i.e. the geometry is still determined by the Euclidean metric g_3.

The Newtonian potential U is related to the field-generating mass density ρ via *Poisson's equation*:†

$$\Delta U = -4\pi G\rho \ , \tag{2.1.20}$$

which can be written in Newtonian coordinates also as

$$R^i{}_{0i0} = -U_{,ii} = 4\pi G\rho$$

or

$$R_{00} = 4\pi G \, \rho \ . \tag{2.1.21}$$

---

† Usually the Newtonian potential is denoted with the opposite sign.

Here,

$$R_{\mu\nu} = R^{\sigma}{}_{\mu\sigma\nu} = \Gamma^{\sigma}_{\nu\mu,\sigma} - \Gamma^{\sigma}_{\sigma\mu,\nu} + \Gamma^{\kappa}_{\nu\mu}\Gamma^{\sigma}_{\sigma\kappa} - \Gamma^{\kappa}_{\sigma\mu}\Gamma^{\sigma}_{\nu\kappa} \qquad (2.1.22)$$

are the components of the *Ricci tensor*, non-vanishing for $\mu = \nu = 0$. In this sense the Newtonian field equations of gravity relate the field generating mass density to the Ricci tensor of the Newtonian space-time. For *isolated systems* the boundary condition

$$\lim_{\substack{|\mathbf{x}| \to \infty \\ t=\text{const.}}} U(t,\mathbf{x}) = 0 \qquad (2.1.23)$$

then selects a unique solution for U:

$$U(t,\mathbf{x}) = \int \frac{\rho(t,\mathbf{x}')}{|\mathbf{x} - \mathbf{x}'|} \, d^3x' \ . \qquad (2.1.24)$$

2.2 Relativistic Space-Time

At this point we would like to ask how the Newtonian theory of gravity differs from a relativistic one like Einstein's theory. Expressed in a simplified manner

— in a relativistic theory of gravity there exists a Lorentzian *space-time metric g*, i.e. a spatial temporal length element

$$ds^2 = g_{\mu\nu}\, dx^\mu dx^\nu \quad ; \quad \mu,\nu = 0,1,2,3 \ ,$$

from which the connection and the curvature tensor are determined uniquely in the frame of Riemannian geometry, i.e.

$$\Gamma^{\mu}_{\nu\lambda} = \frac{1}{2} g^{\mu\sigma}(g_{\sigma\nu,\lambda} + g_{\sigma\lambda,\nu} - g_{\nu\lambda,\sigma}) \qquad \text{(Christoffel symbols)} \qquad (2.2.1)$$

— at each point there exist local coordinates, such that g can be written as

$$g = \eta_{\mu\nu}\, dx^\mu dx^\nu \qquad (2.2.2)$$

with

$$\eta_{\mu\nu} = \text{diag}(-1,+1,+1,+1) \qquad (2.2.3)$$

(local Minkowski-space structure).

Such a space-time metric does not exist in the Newtonian space-time because of the Newtonian absolute time fibration; the Newtonian curvature tensor therefore does *not* exhibit the symmetry condition:†

$$R_{\mu\nu\lambda\sigma} = -R_{\nu\mu\lambda\sigma} \ , \qquad (2.2.4)$$

---

† It is possible, and sometimes useful, to introduce a pair of degenerate metrics into the space-time formulation of Newton's theory. See, e.g., Ehlers (1981).

that might be understood as the condition of integrability of a Riemannian connection to be metric. This, however, is changed if the results from special relativity (Minkowski-space theory) are taken into account. In Newtonian space-time one assumes the availability of an arbitrarily fast signal by means of which arbitrarily remote clocks can be synchronized to show Newtonian absolute time. The structure of Minkowski-space is determined by the fact that no signal can propagate faster than the speed of light. This leads to the relativity of the notion of simultaneity. Formally, the structure of this Minkowski space-time can be expressed very simply: it refers to a flat Riemannian space with indefinite metric g (signature $-+++$), that determines the propagation of light signals as well as the geodetic motion of test bodies. Light signals determine the light cone where

$$ds^2 = 0 \tag{2.2.5}$$

and the classification of vectors into timelike $((v, v) < 0)$, lightlike $((v, v) = 0)$ and spacelike $((v, v) > 0)$, where

$$(u, v) = g_{\mu\nu} u^\mu v^\nu \ . \tag{2.2.6}$$

Since the curvature tensor vanishes in Minkowski-space there exists a preferred class of reference frames with induced coordinates (see Chapter 3) such that the metric tensor takes the form

$$g_{\mu\nu} = \eta_{\mu\nu} = \text{diag}(-1, +1, +1, +1) \ . \tag{2.2.7}$$

This is the class of inertial frames whose representatives are related by elements of the Poincaré group that replaces the Galilean group of Newtonian space-time. Instead of (2.1.5) we now have

$$x'^\mu = \Lambda^\mu_\nu x^\nu + a^\mu \ , \tag{2.2.8}$$

where $\{\Lambda^\mu_\nu\}$ contains the set of velocity transformations (proper Lorentz transformations):

$$\Lambda^i_j = \delta^i_j + v_i v_j \frac{(\gamma' - 1)}{\mathbf{v}^2} \quad ; \quad \Lambda^0_0 = \gamma'$$

$$\Lambda^0_i = -\gamma' v_i/c \quad ; \quad \Lambda^i_0 = -\gamma' v_i/c, \tag{2.2.9}$$

where

$$\gamma' = (1 - \beta'^2)^{-1/2} \quad ; \quad \beta' = \mathbf{v}/c \tag{2.2.10}$$

and c denotes the value of the speed of light in vacuum.

A theory of gravity is called *metric* on a (4-dimensional) space-time $\mathcal{M}$, if (e.g. Will 1981):

— $\mathcal{M}$ is endowed with a metric g

— the worldlines of uncharged test bodies are geodesics w.r.t. g

— in a local, freely falling system all non-gravitational laws of physics take their usual form in Minkowski-space.

The validity of *Einstein's equivalence principle* (e.g. Will 1981) then has the consequence that, in contrast to Newton's theory, each relativistic theory of gravity can be formulated as a metric theory.

Einstein's equivalence principle. *The weak equivalence principle is valid. Furthermore, the outcome of any local, non-gravitational test experiment is independent of the velocity of the freely falling apparatus w.r.t. the fixed stars and where and when it is performed.*

Einstein's theory of gravity is the simplest of all metric theories of the gravitational field. Newton's field equation (2.1.21) is replaced by the *Einstein field equations* $(T = T^\mu_\mu)$:

$$R_{\mu\nu} = C \left(T_{\mu\nu} - \tfrac{1}{2} g_{\mu\nu} T \right) \tag{2.2.11a}$$

or

$$G_{\mu\nu} \equiv R_{\mu\nu} - \tfrac{1}{2} g_{\mu\nu} R = C T_{\mu\nu} \ . \tag{2.2.11b}$$

Here, the field-generating density has been replaced by the symmetric and divergenceless energy-momentum tensor $(T_{\mu\nu})$. The left hand side of (2.2.11b) has been chosen such that the *Einstein tensor* $(G_{\mu\nu})$ does not contain higher than second derivatives and is divergence-free. In Newtonian approximation only $T_{00} = \rho \simeq -T$ contributes; therefore $C(T_{00} - \tfrac{1}{2} g_{00} T) \simeq \tfrac{1}{2} C \rho$ and a comparison with (2.1.21) shows that

$$C = 8\pi G = \frac{8\pi G}{c^4} \ . \tag{2.2.12}$$

Einstein's field equations (2.2.11) relate the field-generating sources in the energy-momentum tensor with the metric g. Thereby the metric tensor is determined up to four coordinate transformations.

In contradistinction to the Newtonian case, where the simple and unique fall-off condition (2.1.23) was sufficient to determine a physically reasonable solution (2.1.24) of the Poisson equation uniquely, the formulation of adequate supplementary boundary conditions to ensure the existence and uniqueness of reasonable solutions of Einstein's field equations, even in the case of 'isolated systems', is still unclear (Damour 1987a and references quoted therein, Ashtekar 1977).

Since in Newtonian approximation R_{00} is given by

$$R_{00} \simeq -\tfrac{1}{2} \Delta g_{00} \tag{2.2.13}$$

the Newtonian potential U is contained in g_{00} according to

$$g_{00} = -1 + 2U/c^2 + \mathcal{O}(c^{-4}) \ . \qquad (2.2.14)$$

It is remarkable that a wealth of exact solutions of the complicated non-linear Einstein equations e.g. for $T_{\mu\nu} = 0$ became known (see e.g. Kramer et al. 1980). However, most of them are of little use, since corresponding connections with field sources are unclear. One exception is the *Schwarzschild metric*, that in standard coordinates (t, r, θ, ϕ) takes the form $(m = GM/c^2)$:

$$ds^2 = -\left(1 - \frac{2m}{r}\right) c^2 dt^2 + \left(1 - \frac{2m}{r}\right)^{-1} dr^2 + r^2(d\theta^2 + \sin^2\theta \, d\phi^2) \quad (2.2.15)$$

and represents the vacuum solution in the exterior space of a spherically symmetric isolated matter distribution or a non-rotating black hole. Already for the *Kerr solution* (e.g. Misner et al. 1973), describing the outer field of a *rotating black hole* it is questionable whether it can be interpreted as the outer solution of some noncollapsed body since there the rotational velocity and the oblateness (or quadrupole moment) of the body are correlated via the equation of state.

For that reason one faces the necessity to consider approximate solutions of Einstein's or competitive field equations. For the analysis of dynamical processes in the solar system this is usually accomplished in the frame of the so-called *(parametrized) post-Newtonian* approximation†. This theory makes use of the fact that the gravitational potential is small in the solar system $(U/c^2 \lesssim 10^{-5})$ and for velocities v, pressure p and internal specific energy density Π one finds

$$\left(\frac{v}{c}\right)^2 , \ \frac{p}{\rho c^2} , \ \Pi \ \lesssim U/c^2 \ .$$

One therefore expands the metrical coefficients in terms of a smallness parameter ϵ with:

$$\epsilon^2 = \mathcal{O}(2) = \left(\frac{v}{c}\right)^2 \sim U/c^2 \sim \frac{p}{\rho c^2} \sim \Pi \qquad (2.2.16)$$

to post-Newtonian order, i.e.

— g_{00} to 4-th, g_{0i} to 3-rd and g_{ij} to 2-nd order,

   if one discusses the motion of massive bodies,

whilst

— g_{00} and g_{ij} to 2-nd order and $g_{0i} \simeq 0$,

   if one is interested in the propagation of electromagnetic signals.

---

† A precise formulation of the concept "Newtonian limit" has been given in Ehlers (1981, 1986), see also Künzle (1976). One can use that formulation as a basis for post-Newtonian expansions; see Lottermoser (1988).

In the post-Newtonian (PN) theory the use of standard PN coordinates (Will 1981) (not to be confused with Schwarzschild standard coordinates) has proved to be very successful. They are quasi-Cartesian coordinates $(t, \mathbf{x})$ with origin usually in the barycenter of the solar system and the spatial part of the metric (g_{ij}) is diagonal and isotropic. The PN theory simply assumes the *existence* of such a coordinate system *for isolated systems* with asymptotic flatness. Tidal forces due to remote and neglected masses M_S ($R_S \sim GM_S/c^2$) lead to deviations from asymptotic flatness of order $(R_S/D)(x/D)^2$, where D denotes the distance of M_S from the barycenter. If, for example we consider the perturbation of α Centauri ($D \sim 4 \times 10^{13}$ km) on the metric at $\mathbf{x}$, it is of order of $10^{-13} \cdot (x[\text{km}]/4 \times 10^{13})^2$, or roughly 10^{-21} in the vicinity of the Pluto orbit.

The asymptotic structure of isolated PN-systems is very similar to that of Minkowski-space. Of greater importance is the fact that the formulation of the PN theory (as a near zone aproximation, see Section 4.7) strongly rests upon the $3+1$ split of space-time $\mathcal{M}$ into space and time, i.e. upon the foliation of $\mathcal{M}$ into space sections which, however, in constrast to the situation in Newtonian space-time, is induced by the "arbitrarily" selected standard PN coordinates. This implies that, as a rule, in the execution of the PN theory one is not dealing with real 4-vectors, 4-tensors etc., i.e. with geometrical quantities. For example, momentum and angular momentum vectors of an isolated system are usually defined by:

$$p^\mu = \int_\Sigma \Theta^{\mu\nu} \, d^3\Sigma_\nu \quad ; \quad J^{\mu\nu} = 2 \int_\Sigma x^{[\mu}\Theta^{\nu]\lambda} \, d^3\Sigma_\lambda \, , \qquad (2.2.17)$$

where Σ represents a space-like hypersurface of the PN space-sections and $\Theta^{\mu\nu}$ the total PN energy-momentum complex. Note that in the definition of $J^{\mu\nu}$ the PN coordinates even appear explicitly!†

In the parametrized PN (PPN)-theory one aims at covering the PN-approximation of a whole class of metrical theories of gravity by means of PN-parameters. For the structure of the metric tensor one first requires only

— g should contain the correct Newtonian limit (2.2.14)

— $g \to \eta + \mathcal{O}(r)$ for $r \gg GM/c^2$, if r characterizes the "distance" to the isolated system of total mass M

— g_{00}, g_{0i}, g_{ij} should transform as scalar, 3-vector and 3-tensor under spatial rotations

— $\mathbf{x}$ should appear only in the form $(\mathbf{x} - \mathbf{x}')$ to avoid the distinction of a certain point and

---

† It is known (e.g. Dixon 1976), that the socalled *world function* is essential for the geometrical construction of theses quantities.

— *g* should be a simple and instantaneous functional of the undifferentiated matter variables (mass, energy, pressure, velocity etc.) whereby gradients of these quantities are classified as non-simple.

In the usual formulation of the PN theory matter is considered as an ideal fluid, i.e. the anisotropic part of the stress tensor is neglected (one exception in found in Misner et al. 1973). With these requirements the general form of *g* can then be written down where the various PN terms are associated with corresponding PN parameters as factors. Both in the selection of matter functionals and in the choice and arrangement of PPN parameters there is some arbitrariness. The latest version of the PPN theory of Will (1981) contains 10 PN parameters denoted by

$$\gamma, \ \beta, \ \xi, \ \alpha_1, \ \alpha_2, \ \alpha_3, \ \zeta_1, \ \zeta_2, \ \zeta_3, \ \zeta_4 \ . \qquad\qquad (2.2.18)$$

Here γ describes the effect of space curvature and β the amount of nonlinearity of the gravitational field. ξ is related to preferred location, $\alpha_1, \alpha_2, \alpha_3$ to preferred frame (e.g. at rest w.r.t. the cosmic microwave background) effects. $\alpha_3, \zeta_1, \ldots, \zeta_4$ are related to a violation of conservation of total momentum. In Einstein's theory of gravity only γ and β are different from zero and $\gamma = \beta = 1$.

So far no experimental evidence exists that besides γ and β any other parameter should be different from zero. In the following we will work only with the Eddington-Robertson parameters γ and β. PPN-metric, Christoffel symbols and components of the Riemann curvature tensor are given in Section A.1 of the Appendix.]

3. Reference Frames and Astrometry

3.1 Introduction

Geodesy, astrometry, high precision navigation — all these disciplines require a set of local and global reference frames in an operative sense. Now, it is widely appreciated that this represents not merely a practical but equally well a conceptual problem, which becomes even stronger with the necessity to leave the classical Newtonian framework and to work in a relativistic space-time.

In a relativistic space-time the situation is complicated by the fact that Einstein's field equations do not determine the metric field uniquely but only up to four coordinate transformations. Physically this implies that in general the relation between the coordinate picture of space-time events and the outcome of a physical measurement (i.e. an *observable*) is quite a complicated one. Now, ever since Einstein's theory of gravity has been formulated theorists have invented a huge number of different methods to tackle this problem. This is not the place to review at least some of them; it is sufficient to express our opinion that only one method invented so far is able to treat this problem in a consistent and satisfactory manner. The basic idea behind this method is to introduce the chrono-geometrical measuring devices (clock and three spatial reference directions along the observer's worldline) explicitly into the theory. This leads to the observer or tetrad formalism, where the observables are obtained by projecting tensors onto tetrad vectors and thereby are automatically scalars, i.e. coordinate independent.

In Newtonian theory a set of massive objects rigidly connected with one another (possibly on a rectangular grid) and equipped with clocks, properly synchronized showing universal Newtonian absolute time, might serve as a primitive version of a reference frame for many purposes.

In a relativistic theory the introduction of rigidity requires some care to restrict the signal propagation velocity by the velocity of light. We also have to abandon the Newtonian notion of clock synchronization. In a relativistic context these problems have to be treated w.r.t. the metric tensor g of space-time and the dynamics of those particles (bodies etc.) that define the reference frame. For simplicity assume these referential particles to fill a certain portion of space, i.e. assume the worldlines $\gamma(\lambda)$ of these particles to be time-like, future-pointing and completely filling a certain part of space-time. Such a structure is called a *time-like congruence* or simply a *platform*. Each element of a platform is called: *observer*. Kinematically a platform very much resembles a fluid and we can use the theory of fluids to study the physical behaviour of a platform. Primarily, the flow of such a fluid is described by the (normalized) tangent vector field u to the various observers. In coordinate representation

$$u^\mu = \frac{dx^\mu}{d\tau} \tag{3.1.1}$$

is the observer's four-velocity if τ is his proper time, indicated by a co-moving natural atomic clock. We can even identify a platform with its corresponding u field. Since u is time-like and normalized it obeys ($c = 1$):

$$(u, u) = -1 \quad ; \quad g_{\mu\nu}u^\mu u^\nu = -1 \ . \tag{3.1.2}$$

Let γ be an observer, T_p the tangent space of all (contravariant) vectors in $p \in \gamma$, then there is a natural 3+1 (space and time) split of T_p, into vectors lying in the observer's rest frame $\mathcal{R}_u$ and orthogonal to u, and vectors in $\mathcal{T}_u$, in the direction of u. The corresponding projection operators into $\mathcal{R}_u$ and $\mathcal{T}_u$ are given by:

$$P = g + u \otimes u \quad ; \quad P_{\mu\nu} = g_{\mu\nu} + u_\mu u_\nu \tag{3.1.3}$$

$$Q = -u \otimes u \quad ; \quad Q_{\mu\nu} = -u_\mu u_\nu \ . \tag{3.1.4}$$

Each vector $X \in T_p$ can be decomposed according to

$$X = PX + QX \quad ; \quad PX \in \mathcal{R}_u \ , \quad QX \in \mathcal{T}_u \ . \tag{3.1.5}$$

To give an example of how this 3+1 split works, consider the angle of two incident light rays with tangent 4-vectors k and k', as determined by the observer with 4-velocity u. One first has to project the null-vectors k and k' into his space-like rest frame $\mathcal{R}_u$, normalize these projected vectors and finally take the scalar product w.r.t. g. The observed relative angle is then given by:

$$\cos\varphi = \frac{(\bar{k}, \bar{k}')}{|\bar{k}||\bar{k}'|} \quad ; \quad \bar{k} = Pk \quad (\bar{k}^\mu = k^\lambda P_\lambda^\mu) \quad \text{etc.} \tag{3.1.6}$$

Now, in general a platform such as, for example, given by the particles constituting the Earth will have a complicated structure. For studying its physical properties the covariant derivative of u is of great help.

Any velocity gradient $u_{\mu;\nu}$ can be decomposed in the following way (Ehlers 1961, Hawking et al. 1973)

$$u_{\mu;\nu} = \omega_{\mu\nu} + \sigma_{\mu\nu} + \frac{1}{3}\theta P_{\mu\nu} - a_\mu u_\nu \tag{3.1.7}$$

with

$$\omega_{\mu\nu} = P_\mu^\rho P_\nu^\sigma \, u_{[\rho;\sigma]} \tag{3.1.8a}$$

$$\theta_{\mu\nu} = P_\mu^\rho P_\nu^\sigma \, u_{(\rho;\sigma)} \tag{3.1.8b}$$

$$\sigma_{\mu\nu} = \theta_{\mu\nu} - \frac{1}{3}\theta P_{\mu\nu} \tag{3.1.8c}$$

$$\theta = u^\mu{}_{;\mu} \tag{3.1.8d}$$

$$a_\mu = u_{\mu;\nu} u^\nu \ . \tag{3.1.8e}$$

Here a with components a^μ denotes the 4-acceleration of the observer.

It is important to visualize the physical significance of these quantities. To this end it is useful to define the neighbourhood of an observer in a platform. Consider a platform $\mathcal{P}$ and a one-parameter set of curves $\mathcal{P}^\perp$ perpendicular to the world-lines of $\mathcal{P}$ (Fig. 3.1).

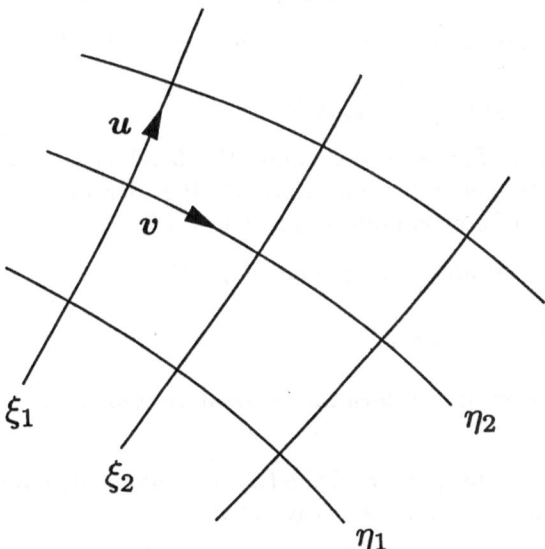

Fig. 3.1. A two-parameter field $x^\mu(\xi, \eta)$ of curves with commuting tangent vectors $u^\mu = \partial x^\mu/\partial\eta$ and $v^\mu = \partial x^\mu/\partial\xi$

Let $u^\mu = \partial x^\mu/\partial\eta$ and let $v^\mu = \partial x^\mu/\partial\xi$ tangent to the $\mathcal{P}^\perp$-curves, then the vectors u and v commute, i.e.

$$[u, v] = \mathcal{L}_u v = (u^\mu v^\nu{}_{,\mu} - v^\mu u^\nu{}_{,\mu})\frac{\partial}{\partial x^\nu}$$

$$= \left(\frac{\partial^2 x^\nu}{\partial\xi\partial\eta} - \frac{\partial^2 x^\nu}{\partial\eta\partial\xi}\right)\frac{\partial}{\partial x^\nu} = 0 \ ,$$

where $\mathcal{L}$ denotes the Lie derivative.

Definition. *If v is a vector field with $\mathcal{L}_u v = 0$, its projection into $\mathcal{R}_u$, $n = Pv$ is called a neighbour of u or $\gamma(\eta)$.*

Let n be a neighbour of u, $(n \in \mathcal{R}_u)$ and $e_{(i)}$ three orthonormal spacelike vectors. Then we can write

$$n^\mu = n^{(i)} e^\mu_{(i)} \equiv \mathbf{n} \cdot e^\mu \tag{3.1.9}$$

in obvious notation. **n** might be thought of as representing the position vector of a neighbouring observer defined in his rest frame.

A suitable covariant derivative then will give us a relativistic 3-velocity of **n**, if it has the following property: if $X \in \mathcal{R}_u$ along a worldline γ (i.e. $(X, u) = 0$), then this derivative of X should always lie in $\mathcal{R}_u$. This, however, is just the property of the *Fermi-derivative F_u*, defined by (e.g. Straumann 1984):

$$F_u X = P D_u X = D_u X + (u, X)a - (a, X)u \ . \tag{3.1.10}$$

Again D_u denotes the covariant derivative w.r.t. u, which in a local chart is given by:

$$(D_u)X^\mu = X^\mu_{;\nu} u^\nu = (X^\mu_{,\nu} + \Gamma^\mu_{\nu\sigma} X^\sigma)u^\mu \ . \tag{3.1.11}$$

Therefore, if n is a neighbour of u, $F_u n \in \mathcal{R}_u$ is called the 3-velocity of n w.r.t. γ. $F_u n$ is just the relativistic generalization of dn/dt, if **n** denotes the Newtonian relative position vector of a neighbouring particle w.r.t. γ.

Lemma. *Let $\mathcal{P}$ be a platform, $\gamma \in \mathcal{P}$ and n neighbour of γ. Then*

$$F_u n = D_n u \ ,$$

i.e. the neighbour's 3-velocity is given by the covariant derivative of u w.r.t. n (Sachs et al. 1977).

Physically, this result means that the properties of the (hydrodynamical) flow in the vicinity of γ are given by the covariant derivative of u.

Definition. *A platform $\mathcal{P}$ is called*

i) *geodesic, iff† $a = D_u u = 0$ $(a^\mu = 0)$*

ii) *irrotational, iff $\omega_{\mu\nu} = 0$*

iii) *rigid (in the sense of Born), if $\sigma_{\mu\nu} = \theta = 0$*

iv) *Einstein-synchronizable iff there are ∂‡ functions h and $t = x^0$ such that*

$$u_\mu = g_{\mu\nu} \frac{dx^\nu}{d\tau} = (-h, 0, 0, 0) \ ; \quad \omega_u = -h\, dt$$

---

† iff stands for: if and only if.

‡ $\partial \equiv$ differentiable; in general C^∞.

v) *proper time Einstein-synchronizable if there is a C^1 function t with*

$$u_\mu = (-1, 0, 0, 0) \quad ; \quad \omega_u = -dt$$

vi) *locally proper time Einstein-synchronizable iff*

$$u_{\mu,\nu} - u_{\nu,\mu} = 0 \quad ; \quad d\omega_u = 0 .$$

Thus a platform is geodesic iff all its observers are freely falling and thus each observer $\gamma(\lambda)$ is a geodesic in space-time. According to the Lemma above a platform is irrotational along γ if the neighbours of γ do not rotate relative to axes whose Fermi derivative vanishes along γ. The physical reason for this notation will become more apparent later (Section 3.3.3). A platform is Born rigid along $\gamma(\lambda)$ if the individual neighbours of γ do not change their lengths with λ or time, i.e. if the spatial distance between neighbouring particles remains constant.

If $\mathcal{P}$ is proper time synchronizable any function t with $\omega_u = -dt$ ($u_\mu = -\delta_\mu^0$) is called a proper time function for $\mathcal{P}$ and is uniquely determined up to an additive constant.

Geometrically the relation $\omega_u = -dt$ implies that the $t = $ const. space-like hypersurfaces Σ_t are orthogonal to the u-field. If all observers of $\mathcal{P}$ have set their atomic clocks when their world-lines intersect some Σ_{t_0}, then all clocks will show the same proper time when their world-lines intersect the surface $\Sigma_{t>t_0}$. The Σ_t surfaces ($t = $ const.) therefore are surfaces of proper time simultaneity. If $\mathcal{P}$ is proper time synchronizable, then the clocks of the observers can be synchronized on the sheets of simultaneity according to the *Einstein synchronization prescription* by radar or light signals (Fig. 3.2): let γ and γ' be sufficiently nearby observers of $\mathcal{P}$ such there is a unique light ray starting at a point $p^- \in \gamma$ and hitting γ' in some point p' and a unique light ray connection back from p' to a point $p^+ \in \gamma$. If the atomic clock at γ shows proper time τ^- at p^- and τ^+ at p^+, and the atomic clock at p' the time τ', then the clocks of the two observers are synchronized in the sense of Einstein if:

$$\tau' = \frac{1}{2}(\tau^- + \tau^+) . \tag{3.1.12}$$

Lemma. *(Sachs et al. 1977) A platform is locally proper time Einstein-synchronizable if it is irrotational and geodesic.*

To illustrate this point let us apply Einstein's synchronization procedure in some suitably chosen coordinate system $(t, \mathbf{x})$ (Landau et al. 1962), where the electromagnetic signal propagation is given by:

$$ds^2 = 0 = g_{00} (dx^0)^2 + 2g_{0i} dx^0 dx^i + g_{ij} dx^i dx^j$$

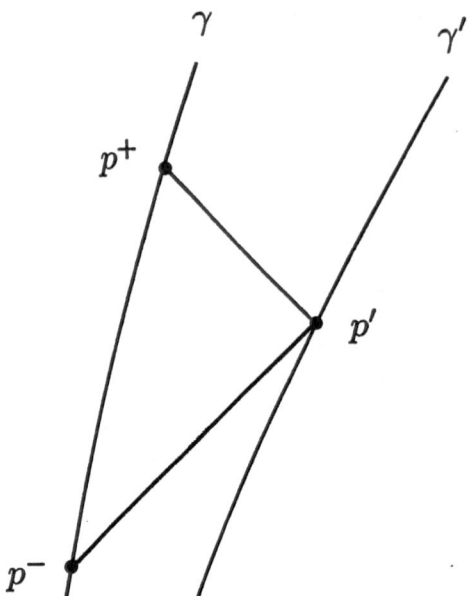

Fig. 3.2. Einstein's synchronization procedure

and the platform is given by $\mathbf{x} = \text{const.}$ Solving this equation for dx^0 gives

$$dx^0_\pm = \frac{1}{g_{00}}\left(-g_{0i}\,dx^i \pm \sqrt{(g_{0i}g_{0j} - g_{ij}g_{00})dx^i dx^j}\right) \ . \tag{3.1.13}$$

Hence, the event (x^0, x^i) on γ is simultaneous to the event $(\bar{x}^0, x^i + dx^i)$ on γ' where

$$\bar{x}^0 = \frac{1}{2}\left[(x^0 + dx^0_-) + (x^0 + dx^0_+)\right] = x^0 - \frac{g_{0i}\,dx^i}{g_{00}} \ . \tag{3.1.14}$$

We see that if $g_{0i} \neq 0$ in our coordinate system clocks cannot be synchronized in the Einstein sense in a unique way. Only if the platform is irrotational and g_{0i} vanishes in a suitable induced coordinate system adopted to the platform are the clocks Einstein-synchronizable.

This has important consequences for the establishment of a consistent and useful terrestrial reference frame: it implies that clocks cannot be synchronized in the sense of Einstein on the rotating Earth and hence a useful concept of a world-time serving many referential purposes cannot simply use the readings of a set of atomic clocks fixed to the Earth's surface.

A platform might be considered a rudimentary version of a reference frame. In the definition of a reference frame we require in addition the existence of three orthonormal vector-fields that lie in the rest space $\mathcal{R}_u$ of $\mathcal{P}$ and serve as spatial reference directions.

Definition. *A platform $\mathcal{P}$ is called a reference frame if there exist three orthonormal spacelike $\textit{\o}$ vector-fields $e_{(i)}$ $(i = 1, 2, 3)$ such that an observer in $\mathcal{P}$ carries a bundle of $\textit{\o}$ orthonormal tetrads $\{e_{(\alpha)}, \alpha = 0, 1, 2, 3\}$ with:*

i) $e_{(0)} = u$

ii) $(e_{(\alpha)}, e_{(\beta)}) = \eta_{\alpha\beta} = \mathrm{diag}(-1, +1, +1, +1)$.

A reference frame of a single observer is simply given by a time-like world line γ (the observer) endowed with a bundle of orthonormal and $\textit{\o}$ tetrads $e_{(\alpha)}$ along γ, i.e. given by $\{\gamma; e_{(\alpha)}, \alpha = 0, 1, 2, 3\}$. $e_{(0)} = u$ determines the passage of time as measured by an atomic clock carried by γ $(u = \partial/\partial\tau)$. It therefore describes the *temporal aspect* in contrast to the space-like vectors $e_{(i)}, i = 1, 2, 3$ describing the *spatial aspect* of a reference frame.

3.2 Clocks and the Temporal Aspect of Reference Frames

According to the last Section it is clear that the readings τ of earthbound atomic clocks cannot be used to define a world-time, that will be of greatest value for applications like navigation, VLBI or analyses of arrival time measurements of pulsar pulses etc. Such a world-time function t, therefore, will not be a proper time and hence not be measurable in a *direct* way.

Now, to a good approximation the solar system can be considered as isolated, so there exists a preferred class of quasi-inertial time coordinates t for an asymptotic observer at rest w.r.t. the barycenter of the solar system. It now turns out that to the required accuracy the time variable t is determined already by the "Newtonian part" of the metric ($\epsilon = v/c$):

$$d\tau^2 = \left(1 - \frac{2U}{c^2}\right) dt^2 - \frac{d\mathbf{x}^2}{c^2} + \mathcal{O}(\epsilon^4) \tag{3.2.1}$$

or

$$\frac{d\tau}{dt} = 1 - \frac{U}{c^2} - \frac{1}{2}\left(\frac{\mathbf{v}}{c}\right)^2 + \mathcal{O}(\epsilon^4) \ . \tag{3.2.2}$$

This has the consequence that the relation between proper time τ and t is independent of the choice (gauge) of spatial coordinates to post-Newtonian order. Clocks in the vicinity of the Earth can be synchronized by means of the coordinate time variable t; this synchronization procedure, which is clearly not in the sense of Einstein, is discussed in Section 5.2.

Proper time τ of an atomic clock at rest at the Earth's surface is related to the so-called International Atomic Time (TAI = Temps Atomique International) that is distributed by the Bureau International de l'Heure (BIH) in Paris. The TAI second equals the SI second (in the international system of units) that equals the duration of 9 192 631 770 periods of that radiation that corresponds to the two hyperfine levels of the ground state of the caesium 133 atom. This definition, however, leaves open where on the Earth's surface the SI second should be realized; due to the gravitational redshift, for example, the rate of a frequency normal at the PTB (Brunswick) differs from that of an atomic clock of the National Bureau of Standards (NBS, Boulder, Colorado) by $\sim g\Delta h/c^2 \sim 2 \times 10^{-13} \sim 5.4\mu s/\text{year}$, where Δh denotes the difference in heights above mean sea level, though both normals realize their own (local) SI second. In the 9th session of the CCDS (Comité Consultatif pour la Définition de la Seconde) in 1980 it was recommended that

"Le TAI est une échelle de temps-coordonnée definie dans un repère de référence geocentrique avec comme unité d'échelle la seconde du SI telle qu'elle est realisée sur le géoide en rotation."

This after translation reads:

"TAI is a coordinate time scale defined in a geocentric reference frame with an SI second realized on the rotating geoid."

This CCDS definition was then supported by the committee for mass and weights (CIPM). In resolutions from the IAU dating from 1976 and 1979 (Kaplan 1981) we find the recommendations:

"The unit of this time scale† be a day of 86 400 SI seconds at mean sea level."

Though the expression coordinate time appears explicitly in the CCDS-definition we would like to comprehend TAI as proper time of an atomic clock at mean sea level:‡

$$\tau^* = TAI \ .$$

According to (3.2.2) proper time τ of an arbitrary earthbound clock can be related with $\tau^* = TAI$:

$$\frac{d\tau}{d\tau^*} \simeq 1 - (U - U_{\text{geoid}})/c^2 \simeq 1 + g\,h/c^2 \simeq 1 + h[\text{km}] \times 10^{-13} \ , \qquad (3.2.3)$$

where h denotes the height of the atomic clock above the geoid.

For practical purposes one introduces two further time scales: TDT (terrestrial dynamical time) and TDB (barycentric dynamical time). Here, the word "dynamical" indicates that the time scale is used in ephemerides calculations.

---

† Here it is referred to TDT which differs from TAI, however, only by an additive constant.

‡ The definition of TAI as proper time is somewhat over-simplified. Strictly speaking TAI is defined by some averaging procedure over the proper times of numerous clocks (Guinot 1986).

According to recommendations of the IAU (Kaplan 1981) TDT and TAI should differ only by an additive constant, and for practical purposes

$$TDT = TAI + 32.184\,\mathrm{s} = \tau^* + 32.184\,\mathrm{s} \qquad (3.2.4)$$

was laid down. According to an IAU resolution from August 1976 barycentric dynamical time TDB and TDT should differ only by periodic terms, i.e. one requires

$$\langle TDB \rangle = \langle TDT \rangle \;, \qquad (3.2.5)$$

where the bracket denotes the time average. According to the above we infer that

$$\frac{d(TDT)}{d(TDB)} \simeq 1 - \frac{1}{c^2}\left(U + \frac{1}{2}\mathbf{v}^2\right)_p + \mathcal{O}(\epsilon^4)\;, \qquad (3.2.6)$$

where the quantities U and $\mathbf{v}$ refer to a point on the rotating geoid and the index p denotes the periodic part of the corresponding quantity, i.e.

$$U_p = U - \langle U \rangle \qquad \text{etc.}$$

Here, all bodies in the solar system *apart from the Earth* contribute to the potential U. Furthermore, we find that

$$dr^* = d(TDT) = \left[1 - \frac{U}{c^2} - \frac{1}{2}\left(\frac{\mathbf{v}}{c}\right)^2\right]_{\text{geoid}} dt$$

$$= \left[1 - \frac{1}{c^2}\left(U + \frac{1}{2}\mathbf{v}^2\right)_p - \frac{1}{c^2}\left\langle U + \frac{1}{2}\mathbf{v}\right\rangle\right]_{\text{geoid}} dt$$

$$= d(TDT)\frac{dt}{d(TDB)} - \frac{1}{c^2}\left\langle U + \frac{1}{2}\mathbf{v}\right\rangle_{\text{geoid}} d(TDT)$$

or

$$\frac{d(TDB)}{dt} = 1 - \frac{1}{c^2}\left\langle U + \frac{1}{2}\mathbf{v}^2\right\rangle_{\text{geoid}} \equiv 1 - \kappa_G$$

$$\equiv \alpha_G \simeq 1 - 1.55 \times 10^{-8} = \text{const.} \qquad (3.2.7)$$

This means that the barycentric coordinate time t appearing in (3.2.1) that is indicated by some clock at rest w.r.t. the barycenter in the solar system in the asymptotic field free region, differs from TDB only by a constant factor; TDB is obtained from t only by an artificial rescaling of units. We thus face the following relations:

$$\tau \overset{(3.2.3)}{\longleftrightarrow} \tau^* = TAI \overset{(3.2.4)}{\longleftrightarrow} TDT \overset{(3.2.6)}{\longleftrightarrow} TDB \overset{(3.2.7)}{\longleftrightarrow} t \;.$$

One further useful time scale is the terrestrial coordinate time t_E. It is defined via

$$\frac{dt_E}{d(TDB)} = 1 - \frac{1}{c^2}\left(U + \frac{1}{2}\mathbf{v}^2\right)\bigg|_p\bigg|_{\mathbf{x}_\oplus} + \mathcal{O}(\epsilon^4) \;, \qquad (3.2.8)$$

where U denotes the Newtonian gravitational potential at the geocenter and $\mathbf{v}$ the velocity of the geocenter in the barycentric system.

We would now like to analyze the expression for $d(TDT)/d(TDB)$ in more detail. Since (A = atomic clock; E = geocenter; B = barycenter)

$$U_p = (U_A - U_E)_p + (U_E)_p \qquad (3.2.9)$$

and ($\mathbf{v}_j^i = \mathbf{v}_j - \mathbf{v}_i$):

$$\mathbf{v}_p^2 = \left[(\mathbf{v}_A^E)^2 + 2\mathbf{v}_E^B \cdot \mathbf{v}_A^E + (\mathbf{v}_E^B)^2\right]_p \qquad (3.2.10)$$

one finds (Moyer 1981):

$$TDB - TDT = \Delta T_0 + \frac{1}{c^2}\int_{t_0}^t \left[U_E + \frac{1}{2}(\mathbf{v}_A^E)^2 + \frac{1}{2}(\mathbf{v}_E^B)^2\right]_p dt$$

$$+ \frac{1}{c^2}\int_{t_0}^t (U_A - U_E + \mathbf{v}_E^B \cdot \mathbf{v}_A^E)_p \, dt \;. \qquad (3.2.11)$$

Now, the term $(\mathbf{v}_A^E)_p^2$ is negligibly small and one finds after partial integration of the last term (Moyer 1981, Thomas 1975):

$$TDB - TDT = \Delta T_0 + \frac{1}{c^2}\int_{t_0}^t \left[U_E + \frac{1}{2}(\mathbf{v}_E^B)^2\right]_p dt + \frac{1}{c^2}\mathbf{v}_E^B \cdot \mathbf{x}_A^E\bigg|_{t_0}^t$$

$$= \text{const.} + (TDB - t_E)_p + \frac{1}{c^2}\mathbf{v}_E^B \cdot \mathbf{x}_A^E\bigg|_{t_0}^t \;. \qquad (3.2.12)$$

In this representation the expression for $TDB - TDT$ is determined by two contributions apart from an additive constant. The term $(TDB - t_E)_p$ is determined by the point dynamics in the solar system, the last term by the barycentric velocity of the Earth and the geocentric position of the atomic clock. If we consider two atomic clocks at different points on the Earth's surface then at time TDB

$$(TDB - TDT)_2 - (TDB - TDT)_1 = -(TDT_2 - TDT_1)$$

$$= \frac{1}{c^2}\mathbf{v}_E^B \cdot (\mathbf{x}_2^E - \mathbf{x}_1^E) \;, \quad (3.2.13)$$

as can easily be seen by means of a Lorentz transformation from the barycentric to the geocentric system. Hence, this last term accounts for the different notions of simultaneity in TDB and TDT. Furthermore, one finds that the constant in (3.2.12) is equal for all clocks at the surface of the Earth.

The dominant term in $(TDB - t_E)_p$ is given by:

$$\frac{1}{c^2} \int_{t_0}^{t} \left[U_E + \frac{1}{2}(\mathbf{v}_E^B)^2 \right]_p dt \simeq \frac{1}{c^2} \int_{t_0}^{t} \left[U_\odot(\mathbf{x}_{EM}) + \frac{1}{2}(\mathbf{v}_{EM}^\odot)^2 \right]_p dt$$

$$\simeq \frac{2}{c^2} \sqrt{m_\odot a}\, e \sin E$$

$$\simeq 1.658 \times 10^{-3} (\sec) \sin E , \qquad (3.2.14)$$

where EM denotes the barycenter of the Earth-Moon system and we have assumed EM to move in a Keplerian orbit with eccentricity e, semi-major axis a and eccentric anomaly E. (This annual term in (3.2.14) was first derived by Aoki 1964). To derive the dominant terms of the position dependent part we write

$$\mathbf{v}_E^B \simeq -\mathbf{v}_\odot^{EM} \simeq - \begin{bmatrix} -\sin L \\ \cos L \cos \varepsilon \\ \cos L \sin \varepsilon \end{bmatrix} , \qquad (3.2.15)$$

where L denotes the mean longitude of the Sun and ϵ the mean obliquity of the ecliptic. In cylindrical coordinates $\mathbf{x}_A^E$ can be written as

$$\mathbf{x}_A^E = \begin{bmatrix} u \cos(\theta_M + \lambda) \\ u \sin(\theta_M + \lambda) \\ v \end{bmatrix} , \qquad (3.2.16)$$

where u denotes the distance from the Earth's rotation axis, v the height above the equator, λ the geographic longitude counted in the eastward direction and θ_M

$$\theta_M = UT1 + L - \pi \qquad (3.2.17)$$

.the mean siderial time or Greenwich hour angle w.r.t. the equinox. Therefore

$$\frac{1}{c^2} \mathbf{v}_E^B \cdot \mathbf{x}_A^E \simeq \frac{v_\odot u}{c^2} \left(\frac{1}{2}(1 + \cos \varepsilon) \sin(UT1 + \lambda) \right.$$

$$\left. - \frac{1}{2}(1 - \cos \varepsilon) \sin(UT1 + \lambda + 2L) \right)$$

$$- \frac{v_\odot(\sin \varepsilon)v}{c^2} \cos L$$

$$\simeq 3 \times 10^{-10} u \sin(UT1 + \lambda)$$

$$- 1.4 \times 10^{-11} u \sin(UT1 + \lambda + 2L)$$

$$- 1.3 \times 10^{-10} v \cos L , \qquad (3.2.18)$$

where the amplitudes are given in seconds. Under simplifying assumptions Moyer (1981) has derived an extended approximation formula for $TDB - TDT$.

In vector form it reads (J = Jupiter; S = Saturn):

$$TDB - TDT = \Delta T_0 + \frac{2}{c^2}(\mathbf{v}_{EM}^{\odot} \cdot \mathbf{x}_{EM}^{\odot}) + \frac{1}{c^2}(\mathbf{v}_{EM}^{B} \cdot \mathbf{x}_{E}^{EM})$$

$$+ \frac{1}{c^2}(\mathbf{v}_{E}^{B} \cdot \mathbf{x}_{A}^{E}) + \frac{m_J}{c^2(m_{\odot} + m_J)}(\mathbf{v}_{J}^{\odot} \cdot \mathbf{x}_{J}^{\odot})$$

$$+ \frac{m_S}{c^2(m_{\odot} + m_S)}(\mathbf{v}_{S}^{\odot} \cdot \mathbf{x}_{S}^{\odot}) + \frac{1}{c^2}(\mathbf{v}_{\odot}^{B} \cdot \mathbf{x}_{EM}^{\odot}) \ . \quad (3.2.19)$$

Pulse arrival time measurements from radio pulsars led to the suspicion (Davies et al. 1985), that Moyer's formula is incorrect at the level of a few μsec. Various groups therefore have aimed at analytically improving the $TDB - TDT$ relationship. To this end Hirayama et al. (1986, 1987) and Fairhead et al. (1986) employ the analytical program VSOP82, that had been developed by Bretagnon (1982) to describe the motion of the planets, and the program ELP2000 by Chapront-Touzé et al. (1983) for the lunar motion.† Expression (3.2.8) then was formally integrated and $TDB - t_E$ was represented as a Poisson series of the form

$$TDB - t_E = \sum_{i,j} (\text{amplitude}) \, T^i \sin\left(\sum_k j_k X_k + \phi\right) \ .$$

Here i and j are integers and X_k the (Hill-Brown) fundamental angles for planets and Moon. Up to higher powers in the time variable they are given in Table 3.1.

Table 3.1. Fundamental Angles in Degrees

$Me = 252.251$	+	$149472.675T$	(Mercury)
$V = 181.980$	+	$58517.816T$	(Venus)
$E = 100.4664$	+	$35999.3729T$	(Earth-Moon)
$M = 355.433$	+	$19140.299T$	(Mars)
$J = 34.351$	+	$3034.906T$	(Jupiter)
$S = 50.077$	+	$1222.114T$	(Saturn)
$U = 314.055$	+	$428.467T$	(Uranus)
$N = 304.349$	+	$218.486T$	(Neptune)
$l = 134.963$	+	$477198.868T$	(Moon from Sun)
$D = 297.850$	+	$445267.112T$	(Moon from perigee)

T is measured in Julian Centuries of 36 525 days of 86 400 seconds of Dynamical Time since J2000.0. The Fundamental epoch is J2000.0 = 2000 January 1.5 = JD2451545.0.

---

† The integration constants were taken from DE200.

Table 3.2. $TDB - t_E$

Argument	Ampl. (μs)	Period (y)	ϕ_i (deg)
$\sin(E + \phi)$	1656.675	1.00	-102.9377
$\sin(E - J + \phi)$	22.418	1.09	-179.916
$\sin(2E + \phi)$	13.840	0.50	154.124
$T\sin(E + \phi)$	10.216	1.00	142.985
$\sin(J + \phi)$	4.770	11.86	-8.888
$\sin(E - S + \phi)$	4.677	1.04	-179.995
$\sin(S + \phi)$	2.257	29.46	-92.481
$\sin(4E - 8M + 3J + \phi)$	1.686	1783.39	106.882
$\sin D$	1.555	0.08	0.000
$\sin(2V - 2E + \phi)$	1.277	0.80	-179.893
$\sin(E - 2J + \phi)$	1.193	1.20	177.357
$\sin(V - E + \phi)$	1.115	1.60	0.009
$\sin(2E - 2J + \phi)$	0.794	0.55	0.829
$\sin(2V - 3E + \phi)$	0.600	3.98	90.894
$\sin(E - U + \phi)$	0.495	1.01	179.989
$\sin(2E - 2M + \phi)$	0.486	1.07	179.733
$\sin(E - N + \phi)$	0.468	1.01	-179.998
$\sin(8V - 13E + \phi)$	0.447	238.92	57.379
$\sin(E - 2M + \phi)$	0.435	15.78	139.592
$\sin(U + \phi)$	0.431	84.02	-174.646
$\sin(3V - 4E + \phi)$	0.376	1.14	91.104
$\sin(3V - 5E + \phi)$	0.243	8.10	165.628
$\sin(E - 2S + \phi)$	0.231	1.07	-86.792
$\sin(2E - J + \phi)$	0.204	0.52	81.739
$\sin(3E + \phi)$	0.173	0.33	51.183
$T\sin(2E + \phi)$	0.171	0.50	40.047
$\sin(2E - 3J + \phi)$	0.159	0.57	10.414
$\sin(2E - 4M + \phi)$	0.144	7.89	122.139
$\sin(3V - 3E + \phi)$	0.138	0.53	-179.452
$\sin(N + \phi)$	0.120	164.77	-43.575
$\sin(3E - 4M + \phi)$	0.119	1.15	150.028
$\sin(2J + \phi)$	0.117	5.93	-19.410
$\sin(D - l + \phi)$	0.102	1.13	180.000

Fundamental epoch: J2000.0 = 2000 January 1.5 = JD2451545.0

The mean longitudes of Mercury Me and of Neptune N that do enter into Table 3.2 at that level of accuracy will appear in Section 4.4 where relativistic effects in the motion of the planetary system are discussed. The results in Table 3.2 are from Hirayama et al. (1987). The phase occuring in the argument of the trigonometrical functions was written in the form

$$\text{phase} = 2\pi T/P_i + \phi_i \; .$$

At the level of $0.04\,\mu$sec these results agree with those obtained by Fairhead et al. (1986).

3.3 Spatial Reference Directions

In the last section we have been concerned with the temporal aspect of reference frames; we now would like to turn to the spatial reference directions. In contrast to the problem of time we here face a wealth of different measuring systems of differing accuracy that independently are in a position to yield a spatial reference system. Examples are

— Laser ranging to artificial satellites such as LAGEOS or to the Moon,

— optical observations or radar measurements in the solar system,

— optical positioning of reference stars (catalogue stars),

— VLBI measurements of quasars or extragalactic radio sources,

and the relation of different spatial reference systems defined operatively represents a *practical* problem of tremendous interest. Here, we would like to deal with the *theoretical, conceptual* problems of spatial reference systems in the frame of a relativistic theory. More precisely, we would like to analyze the three cases in more detail where the spatial reference direction is given by i) the direction of the plumbline in the gravity field of the Earth, ii) (radio) light rays from cosmologically remote sources and iii) by gyroscope systems.

3.3.1 Gravitational Compass and Geoid

For an assumed accuracy of $10^{-11}\,g$ the PPN metric of an isolated, quasi rigid model Earth that rotates uniformly about its Z-axis can in geocentric, comoving coordinates (t, X^μ) be written as

$$g_{00} = -1 + 2U/c^2 - 2\beta U^2/c^4 + \Omega^2(X^2 + Y^2)/c^2 \qquad (3.3.1a)$$

$$(g_{0i}) = \mathbf{g} - \mathbf{X} \times \mathbf{\Omega}/c \qquad (3.3.1b)$$

$$g_{ij} = (1 + 2\gamma U/c^2)\delta_{ij} \qquad (3.3.1c)$$

with

$$\mathbf{g} = -(\gamma + 1)\frac{GJ_\oplus \times \mathbf{X}}{c^3 r^3} \quad . \qquad (3.3.2)$$

This metric is simply obtained from (A1.4) by a rotation of coordinates. From this one finds for the frequency ratios of two earthbound clocks (since $d\tau^2 = -g_{00}\, dt^2$ for $d\mathbf{X} = 0$)

$$\frac{f_1}{f_2} = \frac{\sqrt{-g_{00}}|_2}{\sqrt{-g_{00}}|_1} = \frac{1 - U_2^*/c^2}{1 - U_1^*/c^2} \qquad (3.3.3)$$

with

$$U^* = U + \frac{1}{2}\Omega^2(X^2 + Y^2) - \frac{(2\beta - 1)}{2}\frac{U^2}{c^2} \quad . \qquad (3.3.4)$$

U^* might be considered as *one possible* candidate for a post-Newtonian geopotential and a corresponding geoid might be defined as that $U^* = $ const. surface that lies closest to mean sea level. We call such a distinguished surface of constant gravitational redshift u-geoid, u referring to the 4-velocity $\boldsymbol{u}$ of an atomic clock at rest on the rotating geoid. The realizability of such a u-geoid depends upon the long term stability s_{LT} of atomic clocks; the uncertainty in height in the determination of the geoid is roughly given by $\Delta h \sim (s_{LT}/10^{-14}) \cdot 100\,\mathrm{m}$. At present, for hydrogen masers $s_{LT} \sim 10^{-14}$. However, there already have been plans for a network of clocks for VLBI purposes with phase locking using a geostationary satellite link with long term stability of better than 10^{-15} (Knowles et al. 1982).

On the other hand a relativistic geoid can equally well be defined in the more classical sense by means of gravimetric measurements. Neglecting deformations of the Earth that might be decribed in the usual Newtonian manner, the gravimeter's test mass essentially maintains a constant "distance" from the geocenter by means of a 4-acceleration $\boldsymbol{a}$, pointing away from the geocenter,

$$ \boldsymbol{a} = \frac{D\boldsymbol{u}}{D\tau} \quad ; \quad a^\mu = u^\mu_{;\nu} u^\nu \ . \tag{3.3.5} $$

This 4-acceleration might be used to define the local direction of the plumb-line (unit vector $\boldsymbol{e}_{(a)} = -\boldsymbol{a}/|\boldsymbol{a}|$) and corresponding level surfaces. The condition to keep the test mass "at rest" w.r.t. the geocenter (more precisely w.r.t. the relevant parts of the gravimeter) can be formulated easily if one restricts consideration to an axisymmetrical isolated model Earth rotating about its symmetry axis. In that case the outer gravitational field is stationary and the test particle can be considered at rest in the co-rotating post-Newtonian coordinates X^μ, where the metric is given by (3.3.1). In these coordinates the 4-velocity and 4-acceleration of a co-moving test body are given by:

$$ u^\mu = \frac{dX^\mu}{d\tau} = (c\frac{dt}{d\tau}, \boldsymbol{0}) = \frac{c}{\sqrt{-g_{00}}}(1, \boldsymbol{0}) \tag{3.3.6} $$

and

$$ a^\mu = u^\mu_{,\nu} u^\nu + \Gamma^\mu_{\nu\lambda} u^\nu u^\lambda = (g_{0i} a^i, -\nabla U') \tag{3.3.7} $$

with

$$ U' = U + \frac{1}{2}\Omega^2(X^2 + Y^2) - (\beta + \gamma - 1)\frac{U^2}{c^2} \ . \tag{3.3.8} $$

Now, usually a gravimeter is oriented in accordance with the direction of the plumb-line so the amplitude of gravimetric measurements will be given by

$$ \eta_G = |(\boldsymbol{e}_{(a)}, \boldsymbol{a})| = |(\boldsymbol{a}, \boldsymbol{a})|^{1/2} \ , \tag{3.3.9} $$

where (,) denotes the scalar product w.r.t. the complete metric.† One finds:

$$ \eta_G^2 = (\boldsymbol{a}, \boldsymbol{a}) = (-\nabla \tilde{U})^2 \ , \tag{3.3.10} $$

---

† Note, that Will (1971, 1981) computes only the "radial" acceleration of the gravimetric test particle that usually is not measured.

where the square involves the Euclidean product and

$$\tilde{U} = U + \frac{1}{2}\Omega^2(X^2 + Y^2) - \frac{(2\beta + \gamma - 2)}{2}\frac{U^2}{c^2} \ . \tag{3.3.11}$$

It is interesting to see how the special combination of PPN parameters in the expression for $\tilde{U}$ comes about. To this end we consider a test body in "vertical" free fall with small velocities $v/c \sim 10^{-8}$ for a falling height of $\sim 1\,\mathrm{m}$ and body initially at rest. Without centrifugal, Coriolis and gyroscopic terms, the post-Newtonian equation of motion then simply reads:

$$\ddot{Z} = -\frac{\partial}{\partial Z}U\left(1 - 2(\beta + \gamma)\frac{U}{c^2}\right) \simeq g_0' + aZ \ , \tag{3.3.12}$$

where

$$g_0' = g_0\left(1 - \frac{2(\beta + \gamma)U_0}{c^2}\right) = -\frac{\partial U}{\partial Z}\bigg|_0\left(1 - \frac{2(\beta + \gamma)U_0}{c^2}\right)$$

and the index refers to the initial point of the fall ($a = \mathrm{const.} \simeq 0.3086\,\mathrm{mgal/m}$). Writing the proper falling height ΔZ_p as $\Delta Z_p \simeq (1 + \gamma U_0/c^2)\Delta Z$, the proper falling time interval $\Delta\tau$ as $\Delta\tau \simeq (1 - U_0/c^2)\Delta t$ the solution of (3.3.12) is to sufficient accuracy given by

$$\Delta Z_p \simeq \frac{1}{2}g\,(\Delta\tau)^2\left(1 + \frac{a}{12}(\Delta\tau)^2 + \cdots\right) \ , \tag{3.3.13}$$

where

$$g = g_0\left(1 - (2\beta + \gamma - 2)\frac{U_0}{c^2}\right) \tag{3.3.14}$$

is determined by the expression for $\tilde{U}$.

We now proceed by defining the bundle of a-level surfaces by the following requirements: if $\boldsymbol{b}$ denotes a vector in such a surface, we require: i) $\boldsymbol{b}$ to lie in the rest-space of a co-rotating observer, i.e.

$$(\boldsymbol{b}, \boldsymbol{u}) = g_{\mu\nu}\,b^\mu u^\nu = 0 \tag{3.3.15}$$

and ii) $\boldsymbol{b}$ to be orthogonal to the direction of the plumb-line, i.e. orthogonal to $\boldsymbol{a}$

$$(\boldsymbol{b}, \boldsymbol{a}) = g_{\mu\nu}\,b^\mu a^\nu = 0 \ . \tag{3.3.16}$$

Since $\boldsymbol{u}$ and $\boldsymbol{a}$ are always perpendicular to each other, $(\boldsymbol{u}, \boldsymbol{a}) = 0$, we see that the two conditions i) and ii) uniquely define a two-dimensional level surface.

Theorem. *For any stationary metric the a-level surfaces defined by the conditions (3.3.15) and (3.3.16) above coincide with the u-level surfaces of constant redshift.*

Proof: Since the metric is assumed to be stationary we can choose a coro-tating coordinate system where the metric coefficients are independent of the time variable x^0. For a corotating test body the 4-velocity is then given by (3.3.6) and

$$a_\mu = \left(0, \tfrac{1}{2}\nabla g_{00}/g_{00}\right) \ . \tag{3.3.17}$$

Condition (3.3.16) then gives

$$0 = b^i a_i = g_{00,i} b^i$$

and

$$g_{00} = \text{const.}$$

for an a-level surface spanned by vectors b (q.e.d.).

Hence, in our case, the geometrical u-level surfaces defined by constant clock rates ($U^* = \text{const.}$) coincide with the a-level surfaces, e.g. described by $\tilde{U} = \text{const.}$

3.3.2 The Stellar Compass

We now would like to turn to the case where the spatial reference frame is determined by means of light rays or radio signals originating from objects at cosmic distance. To this end we want to consider the solar system as being isolated, i.e. we assume space-time to approach flat Minkowski space 'far away' from the solar system but still in a regime where effects from Hubble expansion are not appreciable.

Figure 3.3 shows a representation of a 3-dimensional isolated space-time where a conformal transformation of the metric has brought the various asymptotic ('far away') parts ($I^{\pm,0}$, $\mathcal{J}^{\pm,0}$) of space-time into a finite location. In such a (Penrose) diagram light rays must originate from $\mathcal{J}^-$ (past null infinity) in the distant past and will end in $\mathcal{J}^+$ (future null infinity) in the distant future (e.g. Misner et al. 1973, Hawking et al. 1973). Pedestrians moving with velocity less than the speed of light originate from I^- and will end in I^+. Space-like geodesics both originate and end in I^0. Let P be any point of this space-time $\mathcal{M}$ (Fig. 3.3b) and k a null-vector in P. Then there will be exactly one null-geodesic γ through P with k as tangent that intersects $\mathcal{J}^-$ in a point P^-. Let $\bar{P} \neq P$ be another point in $\mathcal{M}$. Then there will be a unique null-ray $\bar{\gamma}$ through $\bar{P}$ that intersects $\mathcal{J}^-$ in $\bar{P}^-$ and whose tangent in $\bar{P}^-$ is 'parallel' (see below) to that of γ in P^-. In this sense one might think of $\mathcal{J}^-$ as mathematically representing the *celestial sphere* and the trajectories γ and $\bar{\gamma}$ as being two light rays originating from one and the same star if the star (or extragalactic radio source, QSO) is sufficiently remote to reveal no significant parallax or proper motion. Calling tangent vectors to γ Weyl parallel to those of $\bar{\gamma}$ we see that the introduction of $\mathcal{J}^-$ (the celestial sphere) allows one to define a global tele-parallelism of tangent vectors to light rays from one and the same fixed star (Soffel et al. 1985, 1988a). In the original manifold this concept is simply realized by comparing the components of the tangent null-vectors in the

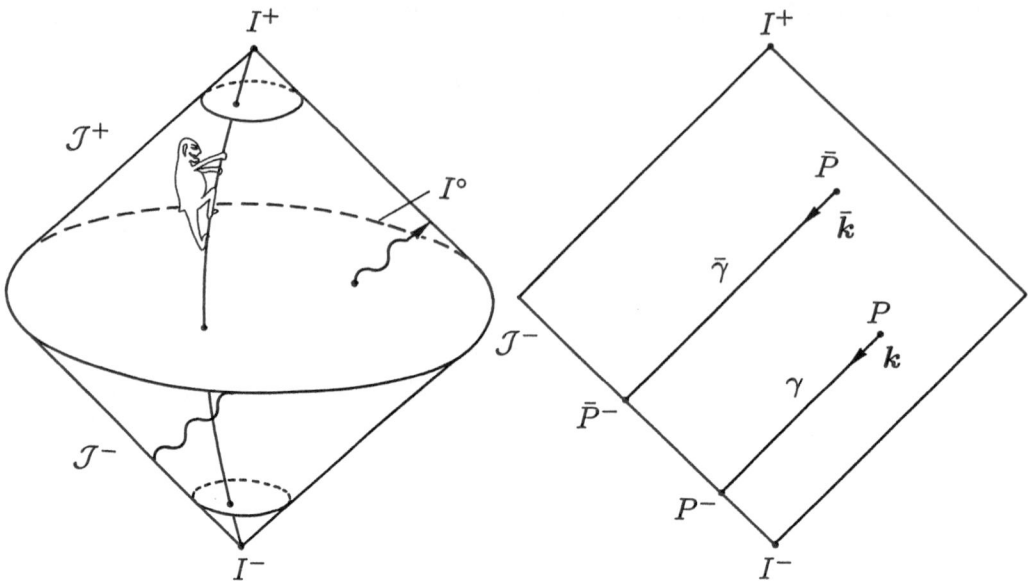

Fig. 3.3. Left (a): schematic representation of a three-dimensional isolated and conformally transformed space-time. Right (b): the geometry in the stellar compass. In these representations, the point at spatial infinity appears as a circle (left) and as a pair of points (right), respectively. For details see, e.g., Ashtekar (1980)

asymptotically flat regime in a suitably chosen coordinate system. Let $(t, \mathbf{x})$ be an asymptotically Cartesian and centered coordinate system (i.e. the "fixed stars at rest" w.r.t. $\mathbf{x}$). Then the two light rays γ and $\bar{\gamma}$ are called "Weyl parallel" (and of the same asymptotic frequency) if $\lim \mathbf{k} = \lim \bar{\mathbf{k}}$ where 'lim' indicates the process of following the light rays back in time and space into the asymptotically flat region. Two Weyl parallel light rays are said to originate from one and the same fixed star, i.e. the bundle of Weyl parallel null-geodesics defines a point on the celestial sphere. Comparing the components of tangent vectors to light rays in the asymptotically flat regime is of course the usual procedure to deal with astrometric problems (Brumberg et al. 1979, Brumberg 1981a, 1981b, Murray 1981, 1983, Pavlov 1984).

We would now like to continue with applications of the concept of such a stellar compass to astrometric measurements. To this end we first notice that for accuracies of $\sim 10^{-3\prime\prime}$ and for directions not directly towards the giant gas planets only the spherical field of the Sun has to be taken into account in post-Newtonian approximation. (In Einstein's theory of gravity the light deflection at the limb of the Sun, due to the post-post-Newtonian contribution of the spherical solar field, due to the oblateness of the Sun and due to the solar angular momentum, amounts to $11 \times 10^{-6\prime\prime}$, $\lesssim 0.2 \times 10^{-6\prime\prime}$ and $0.7 \times 10^{-6\prime\prime}$ respectively (Epstein et al. 1980).)

If we substitute the affine parameter λ by the PPN coordinate time $t = x^0$ (no affine parameter!) in the geodetic equation (2.1.10) by means of

$$\frac{d^2 t}{d\lambda^2} + \Gamma^0_{\nu\sigma} \frac{dx^\nu}{d\lambda} \frac{dx^\sigma}{d\lambda} = 0 \ , \tag{3.3.18}$$

it follows that

$$\frac{d^2 x^i}{dt^2} + \left(\Gamma^i_{\nu\sigma} - \Gamma^0_{\nu\sigma} \frac{dx^i}{dt} \right) \frac{dx^\nu}{dt} \frac{dx^\sigma}{dt} = 0 \ . \tag{3.3.19}$$

The condition for the geodesic to be light-like can be formulated as

$$g_{\mu\nu} \frac{dx^\mu}{dt} \frac{dx^\nu}{dt} = 0 \ . \tag{3.3.20}$$

If therein we substitute the Christoffel symbols from (A1.2), together with

$$\mathbf{x}(t) = \mathbf{x}_0 + \hat{\mathbf{n}}(t - t_0) + \mathbf{x}_p \equiv \mathbf{x}_N + \mathbf{x}_p \ , \tag{3.3.21}$$

we obtain

$$\hat{\mathbf{n}} \cdot \frac{d\mathbf{x}_p}{dt} = -(\gamma + 1)U \tag{3.3.22a}$$

$$\frac{d^2 \mathbf{x}}{dt^2} = (\gamma + 1)\left[\nabla U - 2\hat{\mathbf{n}}\,(\hat{\mathbf{n}} \cdot \nabla U) \right] \ . \tag{3.3.22b}$$

If we now decompose $\mathbf{x}_p$ into components parallel and perpendicular (in the Euclidean sense) to the unperturbed orbit (Will 1981):

$$x_p(t)_\parallel = \hat{\mathbf{n}} \cdot \mathbf{x}_p(t) \tag{3.3.23a}$$
$$\mathbf{x}_p(t)_\perp = (1 - \hat{\mathbf{n}} \otimes \hat{\mathbf{n}})\mathbf{x}_p(t) \tag{3.3.23b}$$

these equations can be written as:

$$\frac{dx_{p\parallel}}{dt} = -(\gamma + 1)U \tag{3.3.24a}$$

$$\frac{d^2 \mathbf{x}_{p\perp}}{dt^2} = (\gamma + 1)\left[\nabla U - \hat{\mathbf{n}}(\hat{\mathbf{n}} \cdot \nabla U) \right] \ . \tag{3.3.24b}$$

If only the spherical field of the Sun is taken into account and the integration of (3.3.24b) is performed along the unperturbed photon orbit with

$$U = \frac{m_\odot}{|\mathbf{x}_0 + \hat{\mathbf{n}}(t - t_0)|} \tag{3.3.25}$$

one obtains

$$\dot{\mathbf{x}}_{p\perp} = -(\gamma + 1)\frac{m_\odot \mathbf{d}}{d^2} \left(\frac{\mathbf{x}_N(t) \cdot \hat{\mathbf{n}}}{r_N(t)} - \frac{\mathbf{x}_0 \cdot \hat{\mathbf{n}}}{r_0} \right) \tag{3.3.26}$$

with

$$\mathbf{d} \equiv \hat{\mathbf{n}} \times (\mathbf{x}_0 \times \hat{\mathbf{n}}) \ . \tag{3.3.27}$$

Together with (3.3.24a) we also find:

$$\dot{\mathbf{x}} = \left(1 - (\gamma+1)\frac{m_\odot}{r}\right)\hat{\mathbf{n}} - (\gamma+1)\frac{m_\odot\mathbf{d}}{d^2}\left(\frac{\mathbf{x}_N \cdot \hat{\mathbf{n}}}{r} - \frac{\mathbf{x}_0 \cdot \hat{\mathbf{n}}}{r_0}\right)$$

$$\simeq \left(1 - (\gamma+1)\frac{m_\odot}{r}\right)\hat{\mathbf{n}} - (\gamma+1)\frac{m_\odot\mathbf{d}}{d^2}(\cos\chi + 1) \ , \tag{3.3.28}$$

where χ denotes the "unperturbed" angle between the directions towards the Sun and the light-source S (Fig. 3.4).

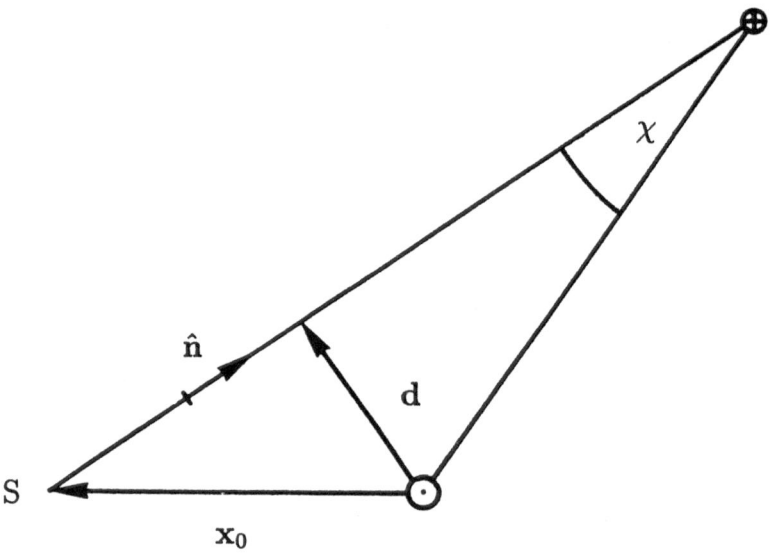

Fig. 3.4. Geometry in the light deflection problem

Further integration of this equation finally gives the PN photon orbit in the spherical field of the Sun as:

$$\mathbf{x}(t) = \mathbf{x}_0 + \hat{\mathbf{n}}(t - t_0)$$
$$- (\gamma+1)m_\odot\left[\hat{\mathbf{n}}\ln\left(\frac{r_N + \mathbf{x}_N \cdot \hat{\mathbf{n}}}{r_0 + \mathbf{x}_0 \cdot \hat{\mathbf{n}}}\right) + \frac{\mathbf{d}}{d^2}[\mathbf{x}_N - r_0 - (t - t_0)]\right]$$

$$\tag{3.3.29}$$

with

$$m_\odot = GM_\odot/c^2 = 1.48 \,\mathrm{km} \ . \tag{3.3.30}$$

Now, two light rays are Weyl parallel if they have a common tangent n̂ in the asymptotic regime that uniquely characterizes the fixed star from which they originate.

According to the presentations of Section 3.1 the angle between two incident light rays with asymptotic directions n̂ and n̂′ as seen by an observer with 4-velocity u is given by

$$\cos\varphi = \frac{(\bar{k}, \bar{k}')}{|\bar{k}||\bar{k}'|}$$

with ($k^\mu = dx^\mu/dt$):

$$\bar{k}^\mu = k^\lambda P_\lambda^\mu \quad ; \quad P_\lambda^\mu = \delta_\lambda^\mu + u_\lambda u^\mu .$$

For an observer with barycentric velocity $\mathbf{v}_\oplus$ we obtain (see also Will 1981, Brumberg 1981) ($\beta = \mathbf{v}_\oplus/c$):

$$\begin{aligned}
\cos\varphi = \hat{\mathbf{n}} \cdot \hat{\mathbf{n}}' &+ (\hat{\mathbf{n}} \cdot \hat{\mathbf{n}}' - 1)\left(\beta \cdot (\hat{\mathbf{n}} + \hat{\mathbf{n}}') + (\hat{\mathbf{n}} \cdot \beta)^2\right. \\
&+ \left.(\hat{\mathbf{n}}' \cdot \beta)^2 + (\hat{\mathbf{n}} \cdot \beta)(\hat{\mathbf{n}}' \cdot \beta) - \beta^2\right) \\
&- (\gamma + 1)\left[\left(\frac{m_\odot}{d}\right)\left(\frac{\mathbf{d} \cdot \hat{\mathbf{n}}'}{d}\right)\left(\frac{\mathbf{x}_\oplus \cdot \hat{\mathbf{n}}}{r_\oplus} - \frac{\mathbf{x}_0 \cdot \hat{\mathbf{n}}}{r_0}\right)\right. \\
&+ \left.\left(\frac{m_\odot}{d'}\right)\left(\frac{\mathbf{d}' \cdot \hat{\mathbf{n}}}{d'}\right)\left(\frac{\mathbf{x}_\oplus \cdot \hat{\mathbf{n}}'}{r_\oplus} - \frac{\mathbf{x}_0' \cdot \hat{\mathbf{n}}'}{r_0'}\right)\right] .
\end{aligned} \tag{3.3.31}$$

Though the theory of astrometric observations can be formulated in a coordinate independent manner it is important to realize that from a practical point of view the construction of a local reference frame using the stellar compass is best achieved by augmenting the Weyl concept with a coordinate dependent construction. If $e_{(i)}$ ($i = 1, 2, 3$) define the fundamental astronomical reference directions (i.e. right ascension α and declination δ) we can write them as

$$e_{(i)} = \left(1 - \gamma\frac{m_\odot}{r}\right)\hat{\mathbf{e}}_{(i)} \tag{3.3.32}$$

if, again, we restrict our discussion to the post-Newtonian field of the Sun. Without aberration effects the observed angle between an incident light ray with tangent vector k and $e_{(i)}$ can then be expressed as

$$\left(e_{(i)}, \frac{\bar{k}}{|\bar{k}|}\right) = \left(1 + (\gamma + 1)\frac{m_\odot}{r}\right)\hat{\mathbf{e}}_{(i)} \cdot \frac{d\mathbf{x}}{dt} ,$$

where $\bar{k}$ denotes the projection of k into the observer's rest space ($u^\mu = dx^\mu/d\tau$):

$$\bar{k}^\mu = (\delta_\lambda^\mu + u_\lambda u^\mu)k^\lambda .$$

This can also be written as

$$\bar{\mathbf{m}} = \mathbf{m} + (\gamma + 1)m_\odot\frac{\mathbf{d} \cdot \hat{\mathbf{e}}}{d^2}(\cos\chi + 1) , \tag{3.3.33}$$

with $\overline{m}_i$ representing the directional cosines of the observed stellar image:

$$\overline{m}_i = - \left(e_{(i)}, \frac{\overline{k}}{|\overline{k}|} \right) \quad ; \quad \overline{m}_x = \cos \overline{\delta} \cos \overline{\alpha} \quad \text{etc.}$$

This immediately leads to the relations

$$\overline{\alpha} - \alpha \simeq \frac{\delta \theta}{\sin \chi} \sec \delta \cos \delta_\odot \sin(\alpha - \alpha_\odot) \tag{3.3.34a}$$

$$\overline{\delta} - \delta \simeq \frac{\delta \theta}{\sin \chi} [\sin \delta \cos \delta_\odot \cos(\alpha - \alpha_\odot) - \sin \delta_\odot \cos \delta] \ , \tag{3.3.34b}$$

with

$$\frac{\delta \theta}{\sin \chi} = (\gamma + 1) \frac{GM_\odot}{c^2 \text{A.U.} \sin^2 \chi} (1 + \cos \chi) \quad ; \quad \frac{GM_\odot}{c^2 \text{A.U.}} \simeq 2'' \times 10^{-3}$$

$$\cos \chi = \sin \delta \sin \delta_\odot + \cos \delta \cos \delta_\odot \cos(\alpha - \alpha_\odot)$$

as recommended by the IAU (Kaplan 1981). One might consider (α, δ) as stellar coordinates in the asymptotic regime. In contrast to the observed angles $(\overline{\alpha}, \overline{\delta})$ they have the great practical advantage of being independent of the solar position in the annual cycle. Such a (coordinate dependent) concept avoids the difficulties that one faces with a purely geometric definition of a local reference frame out of astrometric data.

In the same manner one can introduce reference axes co-moving with the Earth and one finds for the stellar aberration alone:

$$\overline{\mathbf{m}} = \frac{\mathbf{m} + \boldsymbol{\beta}(\gamma' + \sigma)}{\gamma'(1 + \boldsymbol{\beta} \cdot \hat{\mathbf{n}})} \tag{3.3.35}$$

with $(\beta' = v/c)$

$$\gamma' \equiv (1 - \beta'^2)^{-1/2} \quad ; \quad \sigma \equiv \frac{(\gamma' - 1)}{\beta'} \cos \theta \ ,$$

where θ denotes the apex angle (angle w.r.t. the direction of flight) of the stellar image. From this one finds (Stumpff 1979):

$$\tan(\overline{\alpha} - \alpha) = - \frac{\sec \delta (\gamma' + \sigma) \beta_-}{1 + \sec \delta \beta_+ (\gamma' + \sigma)}$$

$$= - \beta_- \sec \delta \left(1 + \frac{1}{2} \beta' \cos \theta - \beta_+ \sec \delta \right) + \mathcal{O}(\beta'^3) \tag{3.3.36a}$$

$$\tan(\overline{\delta} - \delta) = \frac{(\gamma' + \sigma)(\beta_z \cos \delta - \xi \sin \delta)}{1 + (\gamma' + \sigma)(\beta_z \sin \delta + \xi \cos \delta)}$$

$$= (\beta_z \cos \delta - \beta_+ \sin \delta) \left(1 + \frac{1}{2} \beta' \cos \theta - \beta_z \sin \delta - \beta_+ \cos \delta \right)$$

$$- \tan \delta \beta_-^2 + \mathcal{O}(\beta'^3) \tag{3.3.36b}$$

with

$$\xi \equiv \beta_+ - \beta_- \tan\left[\tfrac{1}{2}(\bar{\alpha} - \alpha)\right],$$

$$\beta_+ \equiv \beta'_x \cos\alpha + \beta'_y \sin\alpha,$$

$$\beta_- \equiv \beta'_x \sin\alpha - \beta'_y \cos\alpha .$$

Due to aberration the stellar image appears to be shifted towards the apex of the observer's motion by an angle

$$\Delta\theta = \beta' \sin\theta - \tfrac{1}{4}\beta'^2 \sin 2\theta + \mathcal{O}(\beta'^3) . \tag{3.3.37}$$

For the barycentric motion of the Earth ($\beta'_\oplus \simeq 10^{-4}$) the amplitude of the β'^2 term amounts to about half a milli-arcsecond.

Finally, it should be mentioned that proper motions of astrometrically interesting remote sources (hot spots in quasars etc.) on the celestial sphere can be as large as a few mas. One therefore has to correct for these cosmological effects, for example in VLBI analyzing programs before geodetically relevant information (baselines etc.) can be extracted from astrometric data.

3.3.3 The Inertial Compass

As long as the torques acting on a gyroscope are known to sufficient accuracy the spin-vector S of the gyroscope can be used as spatial reference direction. According to the equivalence principle one expects for a torque free gyro in the co-moving frame where $S = (0, \mathbf{S})$:

$$\frac{d}{dt}\mathbf{S} = 0 . \tag{3.3.38}$$

Furthermore, S should lie in the rest space of the observer, i.e.

$$(S, u) = 0 . \tag{3.3.39}$$

We now want to write the equation of torque free spin motion in a covariant form (Straumann 1984). In a local rest frame we find:

$$D_u S^\mu = S^\mu_{;\nu} u^\nu = \left(\frac{dS^0}{d\tau}, \frac{d\mathbf{S}}{d\tau}\right) = \left(\frac{dS^0}{d\tau}, \mathbf{0}\right) \tag{3.3.40}$$

and from (3.3.39) we infer that

$$(D_u S, u) = -(S, D_u u) = -(S, a) , \tag{3.3.41}$$

where $a = D_u u$ ($a^\mu = u^\mu_{;\nu} u^\nu$) is the 4-acceleration of the observer. Using this result together with (3.3.39) we obtain:

$$(D_u S, u) = -(S, a) = -\left.\frac{dS^0}{d\tau}\right|_R ,$$

where the index refers to the local rest frame and therefore

$$(D_u S)_R = ((S, a), 0)_R = (S, a) \cdot u|_R \ . \tag{3.3.42}$$

Hence, the desired equation of motion for the spin vector reads:

$$D_u S - (S, a)u = 0 \ . \tag{3.3.43}$$

Because of (3.3.39) this result can simply be written as

$$F_u S = D_u S + (u, S)a - (a, S)u = 0 \ , \tag{3.3.44}$$

where F_u denotes the Fermi derivative w.r.t. u.

Definition. *(Fermi frame) A reference frame is called a Fermi frame, if* $F_u e_{(\alpha)} = 0$ *along each observer.*

In this sense the 4-velocity u is always a time-like Fermi axis:

$$F_u u = D_u u + (u, u)a - (a, u)u = 0 \ . \tag{3.3.45}$$

The physical significance of Fermi frames lies in the fact that they represent the "local inertial frames" where Coriolis and centrifugal forces are absent. This can be seen from the form of the *Jacobi equation* describing the acceleration of neighbours in a platform. Let $n = Pv$ be neighbour of u, then (e.g. Hawking et al. 1973):

$$P_\alpha^\gamma \frac{\delta}{\delta s}\left(P_\beta^\alpha \frac{\delta}{\delta s} n^\beta\right) = -R_{\sigma\nu\lambda}^\gamma u^\sigma u^\lambda n^\nu + P_\alpha^\gamma a_{;\nu}^\alpha n^\nu + a^\gamma a_\lambda n^\lambda \ . \tag{3.3.46}$$

In a geodetic Fermi frame ($a = D_u u = 0$) with tetrads $\lambda_{(\alpha)}$ this equation takes a quasi-Newtonian form (see (2.1.16)):

$$\frac{d^2}{ds^2} n^{(i)} = -R^{(i)}{}_{(0)(j)(0)} n^{(j)} \equiv -K^{(i)}{}_{(j)} n^{(j)} \tag{3.3.47}$$

or in compact vectorial notation

$$\frac{d^2}{ds^2} \mathbf{n} + \hat{\mathbf{K}} \mathbf{n} = 0 \tag{3.3.48}$$

with

$$(\hat{\mathbf{K}})_{(i)(j)} = R_{(i)(0)(j)(0)} = R_{\mu\nu\rho\sigma} \lambda_{(i)}^\mu \lambda_{(0)}^\nu \lambda_{(j)}^\rho \lambda_{(0)}^\sigma \ . \tag{3.3.49}$$

Let $e_{(\alpha)}$ be an arbitrary tetrad field along $\gamma(\lambda)$. We can then always write

$$D_u e_{(i)} = \omega_{(i)(j)}\, e_{(j)} + a_{(i)}\, e_{(0)} \ . \tag{3.3.50}$$

The normalization condition of tetrads

$$(e_{(\alpha)}, e_{(\beta)}) = \eta_{\alpha\beta}$$

yields:

$$a_{(i)} = (D_u u, e_{(i)}) = (a, e_{(i)}) \tag{3.3.51}$$

and

$$\omega_{(i)(j)} = -\omega_{(j)(i)} = (D_u e_{(i)}, e_{(j)}) \ . \tag{3.3.52}$$

Together with $e_{(0)} = u$, it now follows that

$$D_u e_{(i)} = (a, e_{(i)})u + \omega_{(i)(j)} e_{(j)} \tag{3.3.53}$$

or

$$F_u e_{(i)} = \omega_{(i)(j)} \, e_{(j)} \ . \tag{3.3.54}$$

For arbitrary axes the Jacobi equation reads

$$\ddot{n} + 2\, \mathbf{\Omega} \times \dot{n} + \mathbf{\Omega} \times (\mathbf{\Omega} \times \mathbf{n}) + \hat{\mathbf{K}} \cdot \mathbf{n} = 0 \ , \tag{3.3.55}$$

where the angular velocity $\mathbf{\Omega}$ is determined by

$$\omega_{(i)(j)} = \epsilon_{(i)(j)(k)} \, \Omega^{(k)} \ . \tag{3.3.56}$$

From the last two relations we see that in a local frame with non-Fermi-transported axes centrifugal and Coriolis forces appear, i.e. the Fermi axes represent indeed the local inertial axes.

We now want to analyze the Fermi transport of the spin vector of a torque free gyroscope in orbit about the Earth in more detail. To this end it is sufficient to consider the linear form of the PPN-metric:

$$ds^2 = (-1 + 2U)dt^2 + (1 + 2\gamma U)d\mathbf{x}^2 + 2g_i \, dx^i dt \tag{3.3.57}$$

with

$$g_i \equiv g_{0i} = -\frac{1}{2}(4\gamma + 3)V_i - \frac{1}{2}W_i$$

in standard PN coordinates. Our strategy now will be the following: we will first analyze the Fermi-transport condition for the *coordinate components* of S. Next, we will construct a *fixed star oriented frame* $e_{(\alpha)}$ co-moving with the observer (cf. Section 3.5) and in the last step we will compute the variations of S w.r.t. fixed star oriented axes which are the objectives in the Stanford gyroscope (GPB) experiment. In our coordinate system the Fermi-transport condition (3.3.44) reads:

$$\frac{dS_\mu}{d\tau} = \Gamma^\lambda_{\mu\nu} S_\lambda u^\nu + g_{\mu\nu} u^\nu a^\lambda S_\lambda \ . \tag{3.3.58}$$

The condition (3.3.39) leads to

$$\frac{dt}{d\tau}S_0 = -\frac{dx^i}{d\tau}S_i$$

or

$$S_0 = -v^i S_i \ . \tag{3.3.59}$$

By the same condition: $a_0 = -v^i a_i$. Therefore,

$$\frac{dS_i}{dt} = \Gamma^j_{i0}S_j - \Gamma^0_{i0}v^j S_j + \Gamma^j_{ik}v^k S_j - \Gamma^0_{ik}v^k v^j S_j$$
$$+ (g_{0i} + g_{ik}v^k)(-a^0 v^j S_j + a^j S_j)$$

or, to third order in v/c,

$$\frac{dS_i}{dt} = \left(\overset{(3)}{\Gamma}{}^j_{i0} - \overset{(2)}{\Gamma}{}^0_{i0}v^j + \overset{(2)}{\Gamma}{}^j_{ik}v^k \right) S_j + v_i \, a^j S_j \ . \tag{3.3.60}$$

Inserting the Christoffel symbols from A.1 of the Appendix we end up with

$$\frac{dS_i}{dt} = -(2\gamma + 2)V_{[j,i]}S_j + \gamma U_{,0} \, S_i + (\gamma + 1)(\mathbf{v} \cdot \mathbf{S})U_{,i}$$
$$+ \gamma S_i(\mathbf{v} \cdot \nabla U) - \gamma v_i(\mathbf{S} \cdot \nabla U) + v_i(\mathbf{a} \cdot \mathbf{S}) \ . \tag{3.3.61}$$

This describes the coordinate motion of the gyro's spin vector. We now want to construct the co-moving fixed star oriented tetrad.† To this end we start with the coordinate induced tetrad (cf. Section 3.4.2), whose covariant components are given by:

$$\tilde{e}_{(0)\mu} = ((1 - U); -g_i) \quad ; \quad \tilde{e}_{(1)\mu} = (0; (1 + \gamma U), 0, 0) \quad \text{etc.} \tag{3.3.62}$$

This represents the local reference frame of an observer at rest in our PN coordinate system whose world line is given by $\mathbf{x} = \text{const.}$ The co-moving fixed star oriented tetrad $e_{(\alpha)}$ is then obtained from $\tilde{e}_{(\alpha)}$ by means of a Lorentz boost such that $e^\mu_{(0)} = u^\mu$. This boost is performed with the gyro's 3-velocity $\tilde{v}^i$ w.r.t. $\tilde{e}_{(\alpha)}$ and *not* with the coordinate velocity v^i:

$$\tilde{v}^i = \frac{\tilde{u}^i}{\tilde{u}^0} = \frac{u^\mu e_{(i)\mu}}{u^\mu e_{(0)\mu}} \simeq \frac{(1 + \gamma U)u^i}{(1 - U)u^0} \simeq (1 + (\gamma + 1)U)v^i \ . \tag{3.3.63}$$

---

† Strictly speaking tetrads at rest w.r.t. the fixed stars exist only in stationary gravitational field with Killing vector-field ξ. There they are defined by $\mathcal{L}_\xi e_{(\alpha)} = 0$ and are called *Copernican tetrads* (e.g. Straumann 1984).

Now, let us denote the components of S w.r.t. $\tilde{e}_{(\alpha)}$ by $\tilde{S}^\mu$ and the components w.r.t. $e_{(\alpha)}$ by S^μ. A Lorentz-transformation from $e_{(\alpha)}$ to $\tilde{e}_{(\alpha)}$ with $-\tilde{v}$ then gives to sufficient accuracy:

$$\tilde{S}^\mu = \left(\tilde{v}_j S^j, S^i + \frac{1}{2}\tilde{v}^i(\tilde{v}_j S^j) \right) \ . \tag{3.3.64}$$

Furthermore, the relation between $\mathcal{S}^i$ and S^i is given by:

$$\mathcal{S}^i = (1 - \gamma U)\tilde{S}^i = (1 - \gamma U)S^i + \frac{1}{2}v^i(v_j S^j)$$

or

$$S_i = (1 + \gamma U)\mathcal{S}_i + \frac{1}{2}v_i(v^j \mathcal{S}_j) \ . \tag{3.3.65}$$

Inverting this relation gives

$$\mathcal{S}_i = (1 - \gamma U)S_i - \frac{1}{2}v_i(v^j S_j) \ . \tag{3.3.66}$$

From this the time derivative of $\mathcal{S}_i$ is obtained as

$$\dot{\mathcal{S}}_i = \dot{S}_i - \gamma S_i(\dot{U} + \mathbf{v} \cdot \nabla U) - \frac{1}{2}\dot{v}_i(v^j S_j) - \frac{1}{2}v_i(\dot{v}^j S_j)$$

or, using $\dot{\mathbf{v}} = \nabla U + \mathbf{a}$,

$$\dot{\mathcal{S}}_i = \dot{S}_i - \gamma S_i(\dot{U} + \mathbf{v} \cdot \nabla U) - \frac{1}{2}U_{,i}(\mathbf{v} \cdot \mathbf{S}) - \frac{1}{2}v_i(\mathbf{S} \cdot \nabla U)$$
$$- \frac{1}{2}a^i(\mathbf{v} \cdot \mathbf{S}) - \frac{1}{2}v^i(\mathbf{a} \cdot \mathbf{S}) \ . \tag{3.3.67}$$

Inserting our result (3.3.61) for $\dot{S}_i$ we finally obtain:

$$\dot{\mathcal{S}}_i = \mathcal{S}_j \left(v_{[i}a_{j]} - (2\gamma + 2)V_{[j,i]} - (2\gamma + 1)v_{[i}U_{,j]} \right) \ . \tag{3.3.68}$$

Since

$$g_{[i,j]} = \zeta_{[i,j]} = -(2\gamma + 2)V_{[i,j]} \tag{3.3.69}$$

the second term in brackets can also be written as

$$g_{[j,i]} = g_{0[j,i]} \ .$$

If $\mathbf{S}$ now denotes the spin components w.r.t. the co-moving fixed star oriented frame ($\mathbf{S} = (\mathcal{S}_1, \mathcal{S}_2, \mathcal{S}_3)$) the final result can be written in the form:

$$\frac{d\mathbf{S}}{d\tau} = \mathbf{\Omega} \times \mathbf{S} \tag{3.3.70}$$

with

$$\boldsymbol{\Omega} = -\tfrac{1}{2}\mathbf{v} \times \mathbf{a} - \tfrac{1}{2}\nabla \times \mathbf{g} + \left(\gamma + \tfrac{1}{2}\right)\mathbf{v} \times \nabla U \ . \tag{3.3.71}$$

The first term describes the *Thomas precession*, the second the *Lense-Thirring* or *Schiff precession* and the last term the *geodetic precession*.

The Thomas precession (Thomas 1927) plays a role in atomic physics because of the electronic spin. Purely kinematic in origin it here results from the Fermi-Walker transport of the electronic spin in course of the electron motion about the nucleus. Its corresponding Thomas interaction energy $U_{\mathrm{Th}} = \mathbf{S} \cdot \boldsymbol{\Omega}_{\mathrm{Th}}$ can be rewritten to take a form similar to that of the usual spin orbit coupling resulting from the magnetic dipole interaction of the electron spin (Jackson 1975):

$$U_{\mathrm{Th}} = \mathbf{S} \cdot \boldsymbol{\Omega}_{\mathrm{Th}} = -\frac{1}{2m^2 c^2}(\mathbf{S} \cdot \mathbf{L})\frac{1}{r}\frac{dV}{dr} \ , \tag{3.3.72}$$

where $\mathbf{L}$ denotes the electronic angular momentum and V the Coulomb potential of the nucleus. The Thomas precession therefore *reduces* the spin orbit interaction energy by a factor of two. Without this additional Thomas energy the anomalous Zeeman effect can be explained with an electronic g-factor of two; however, the fine structure intervals are then wrong by a factor of two. The fine structure levels are described correctly with $g = 1$ but then the anomalous Zeeman effect cannot be explained. In atomic nuclei the Thomas precession of the nucleons results from the nuclear forces that dominate the electromagnetic forces and thus the observed spin orbit splitting in nuclei is determined by Thomas precession. This explains why the doublets from single particle excitation in nuclei are "inverted" (Jackson 1975).

The Lense-Thirring precession results from the gravitomagnetic field of the field generating source. According to our discussion above it implies a "dragging of inertial frames". Our rotating central body influences the surrounding space-time in some sense similar as if it were immersed in a viscous fluid transferring some of its rotational energy to the surrounding medium by frictional forces (Misner et al. 1973). This frame-dragging effect was first discussed by Lense et al. (1918) in the orbital motion of a satellite about a central rotating body (cf. Section 4.2). Schiff (1960; see also Pugh 1959) proposed to determine this frame-dragging effect with the analysis of the spin axes of torque-free gyros in orbit about the Earth.

The geodetic precession was first discussed by de Sitter (1916) in his classic paper on the lunar motion (cf. Section 4.3) and later interpreted as a tetrad effect by Fokker (1921, 1965).

For a gyroscope placed in orbit about the Earth in a "drag-free" satellite $(a^\mu = 0)$

$$\mathbf{g} = -(\gamma + 1)G\frac{\mathbf{J}_\oplus \times \mathbf{x}}{r^3}$$

and the Lense-Thirring precession frequency can be written as

$$\boldsymbol{\Omega}_{\mathrm{LT}} = -\tfrac{1}{2}\nabla \times \mathbf{g} = \left(\frac{\gamma+1}{2}\right)\frac{G}{c^2 r^3}\left[-\mathbf{J}_\oplus + \frac{3(\mathbf{J}_\oplus \cdot \mathbf{x})\mathbf{x}}{r^2}\right] \ . \tag{3.3.73}$$

With $U \simeq GM/r$ the geodetic precession frequency can be written as

$$\mathbf{\Omega}_{GP} = \left(\gamma + \frac{1}{2}\right) \frac{GM}{c^2} \frac{\mathbf{x} \times \mathbf{v}}{r^3} \; . \tag{3.3.74}$$

If we consider a circular geodesic orbit with radius r and unit vector $\mathbf{n}$ normal to the orbital plane with

$$\mathbf{v} = -\left(\frac{GM}{r^3}\right)^{1/2} \mathbf{x} \times \mathbf{n} \; ,$$

the total spin precession frequency for torque-free gyros *averaged* over one complete orbit reads (Weinberg 1972):

$$[\mathbf{\Omega}] = \frac{G}{2r^3} \left(\mathbf{J}_\oplus - \mathbf{n}(\mathbf{n} \cdot \mathbf{J}_\oplus)\right) + 3 \frac{(GM)^{3/2} \mathbf{n}}{2r^{5/2}} \; . \tag{3.3.75}$$

Hence for low flying satellites ($r \sim R_\oplus$):

$$\frac{\text{Lense} - \text{Thirring}}{\text{geodetic}} \simeq \frac{GJ_\oplus}{3(GM)^{3/2} R_\oplus^{1/2}} = 6.5 \times 10^{-3} \; . \tag{3.3.76}$$

How do the Fermi tetrads moving with the barycenter of the Earth-Moon system about the Sun (or the barycenter of the solar system) look in the frame of the PPN-theory? To answer this question we will neglect the planetary influences upon the motion of the Fermi frame and assume the orbit of the Earth-Moon barycenter about the Sun to be circular and geodesic ($\mathbf{a} = 0$). Furthermore, we can neglect the solar angular momentum, i.e. the Lense-Thirring precession. Hence, we want to integrate the parallel transport law for gyroscope axes along the world-line γ_B of the Earth-Moon barycenter under these simplifying assumptions. According to A.2 the world-line γ_B in the equatorial plane ($\theta = \pi/2$) of the PN *polar coordinates* is given by:

$$t = \sigma_{+3/2} \, \tau$$
$$r = a$$
$$\theta = \pi/2$$
$$\phi = n(t - t_0) = n_\tau(\tau - \tau_0) \tag{3.3.77}$$

with

$$n = n_0 \left(1 - \frac{(2\beta + \gamma)}{2} \sigma\right)$$
$$n_\tau = n_0 \left(1 + \frac{(3 - 2\beta - \gamma)}{2} \sigma\right)$$

$$n_0^2 = m_\odot \, a^{-3}$$
$$\sigma = m_\odot/(c^2 a)$$

and the notation

$$\sigma_\epsilon = 1 + \epsilon\sigma \ .$$

Here $\tau = \sigma_{-3/2}\, t$ indicates the proper time of an observer co-moving with γ_B. His 4-velocity is then given by:

$$u^\mu = \frac{dx^\mu}{d\tau} = (\sigma_{+3/2}, 0, 0, n_\tau) \ , \tag{3.3.78}$$

which agrees with the time-like component of the Fermi frame $(e_{(0)} = u)$. For the spherical field of the Sun the Christoffel symbols along γ_B are given by (see (A1.9)):

$$\Gamma^r_{rr} = -\frac{\sigma}{a}\gamma \quad ; \quad \Gamma^t_{rt} = +\frac{\sigma}{a} \quad ; \quad \Gamma^r_{tt} = +\frac{\sigma}{a}\sigma_{-(2\beta+\gamma)} \tag{3.3.79a}$$

$$\Gamma^\theta_{\theta r} = \Gamma^\phi_{\phi r} = -\frac{1}{a^2}\Gamma^r_{\theta\theta} = -\frac{1}{a^2}\Gamma^r_{\phi\phi} = \frac{1}{a}\sigma_{-\gamma} \ . \tag{3.3.79b}$$

The parallel transport law for torque-free spin axes reads:

$$\frac{d\lambda^\mu_{(\alpha)}}{d\tau} + \Gamma^\mu_{\nu\kappa}\,\lambda^\nu_{(0)}\lambda^\kappa_{(\alpha)} = 0 \ , \tag{3.3.80}$$

if we choose the proper time τ as affine parameter. To sufficient accuracy these equations for $\lambda^\mu_{(i)}$ read:

$$\frac{d\lambda^t}{d\tau} + an_0^2\,\lambda^r = 0 \tag{3.3.81a}$$

$$\frac{d\lambda^r}{d\tau} + an_0^2\,\lambda^t - an_\tau\sigma_{-\gamma}\,\lambda^\phi = 0 \tag{3.3.81b}$$

$$\frac{d\lambda^\theta}{d\tau} = 0 \tag{3.3.81c}$$

$$\frac{d\lambda^\phi}{d\tau} + \frac{1}{a}n_\tau\sigma_{-\gamma}\,\lambda^r = 0 \ . \tag{3.3.81d}$$

These can easily be solved by setting

$$\begin{aligned}
\lambda^r_{(r)} &= A\cos\phi_F \\
\lambda^r_{(\theta)} &= 0 \\
\lambda^r_{(\phi)} &= A\sin\phi_F
\end{aligned} \tag{3.3.82}$$

with

$$\phi_F = \alpha\, n_\tau(\tau - \tau_0) \quad ; \quad \alpha \equiv \sigma_{-(2\gamma+1)/2} \ . \tag{3.3.83}$$

We find $A = \sigma_{-\gamma}$ and a possible solution for the Fermi tetrads to order σ^2 reads:

$$
\begin{bmatrix}
\lambda^{\mu}_{(t)} \\
\lambda^{\mu}_{(r)} \\
\lambda^{\mu}_{(\theta)} \\
\lambda^{\mu}_{(\phi)}
\end{bmatrix}
=
\begin{bmatrix}
\sigma_{+3/2}, & 0, & 0, & n_0 \\
- an_0 \sin \phi_F, & \sigma_{-\gamma} \cos \phi_F, & 0, & -\sigma_{(1/2-\gamma)} \sin \phi_F / a \\
0, & 0, & \sigma_{-\gamma}/a, & 0 \\
+ an_0 \cos \phi_F, & \sigma_{-\gamma} \sin \phi_F, & 0, & +\sigma_{(1/2-\gamma)} \cos \phi_F / a
\end{bmatrix}.
$$

$$(3.3.84)$$

This form of Fermi tetrads is not difficult to comprehend. Starting again with the coordinate induced tetrad

$$
e_{(t)} \simeq \sigma_+ \frac{\partial}{\partial t} \quad ; \quad e_{(r)} \simeq \sigma_{-\gamma} \frac{\partial}{\partial r} \quad ; \quad e_{(\theta)} \simeq \frac{1}{a}\sigma_{-\gamma} \frac{\partial}{\partial \theta} \quad ; \quad e_{(\phi)} \simeq \frac{1}{a}\sigma_{-\gamma} \frac{\partial}{\partial \phi}
$$

$$(3.3.85)$$

and changing to the co-moving system by a Lorentz boost in $e_{(\phi)}$-direction and finally rotating the axes about $e_{(\theta)}$ such that the axes are fixed star oriented we obtain the co-moving fixed star oriented tetrad $\bar{e}_{(\alpha)}$, which differs from the Fermi tetrad only by the argument of the trigonometrical functions. Instead of ϕ_F we there find the angle ϕ of the orbital motion, i.e.

$$
\lambda_{(\alpha)} = \bar{e}_{(\alpha)} [\phi \rightarrow \phi_F] \quad ; \quad \phi_F = \phi + \Omega_{\mathrm{GP}} \tau \; ,
$$

$$(3.3.86)$$

where

$$
\Omega_{\mathrm{GP}} = \alpha \, n_r - n_\tau = -\frac{(2\gamma + 1)}{2} \sigma n_0
$$

$$(3.3.87)$$

is the geodetic precession frequency.

We would now like to discuss a gyroscope of a special kind, namely a laser-gyro of Sagnac type. In such an instrument we find two laser beams travelling around a closed circuit and brought to interference (Fig. 3.5).

We first consider such a system in flat space rotating with angular velocity Ω w.r.t. the fixed stars. The metric in the rotating platform is then given by (3.3.1) with $U = 0 = \mathbf{L}$. Assuming $v/c \ll 1$ we write

$$
g_{\mu\nu} = \eta_{\mu\nu} + h_{\mu\nu}
$$

$$(3.3.88)$$

with small functions $h_{\mu\nu}$. We would now like to analyze Maxwell's equations for the laser beams in the eikonal-approximation, where the vector potential A_μ can be written as

$$
A_\mu = a_\mu e^{iS}
$$

$$(3.3.89)$$

with a rapidly varying phase S and

$$
g_{\mu\nu} S_{,\mu} S_{,\nu} = 0 \; .
$$

$$(3.3.90)$$

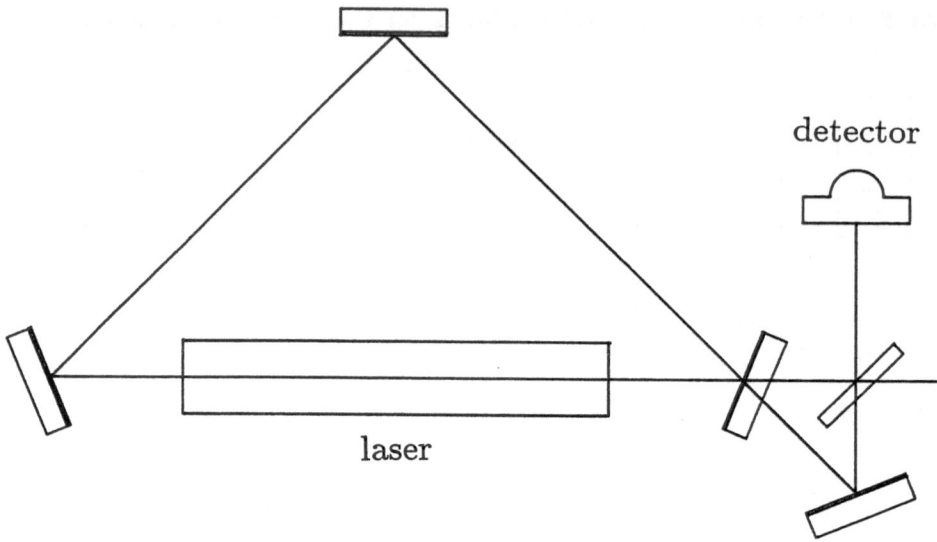

Fig. 3.5. A laser gyroscope of Sagnac type

If the index 0 refers to inertial axes, i.e.

$$k_0^\mu = \left(\frac{\omega}{c}, \mathbf{k}\right), \quad |\mathbf{k}| = \frac{\omega}{c} \quad \text{etc.} , \qquad (3.3.91)$$

then to first order in $h_{\mu\nu}$ Maxwell's equations lead to (Scully et al. 1981):

$$k_0^\mu \,\Delta S_{,\mu} = \frac{1}{2} k_0^\mu k_0^\nu \, h_{\mu\nu}$$

or

$$\Delta S = \frac{1}{2} \frac{\omega}{c} \int h_{\mu\nu} \, n^\mu dx^\nu \qquad (3.3.92)$$

with

$$S = S_0 + \Delta S \quad ; \quad k_0^\mu = \frac{\omega}{c} n^\mu .$$

If we now denote the light ray travelling clockwise (anti-clockwise) by $+(-)$, $n_\pm^\mu = (1, \pm n^i)$ and use the inertial time t as parameter, we obtain:

$$(\Delta S)_\pm = \frac{1}{2} \omega^\pm \int_0^T (h_{00} \pm 2h_{00} n^i + h_{ij} n^i n^j) \, dt$$

and

$$(\Delta S)_{\mathrm{T}} = (\Delta S)_+ - (\Delta S)_- = (\omega^+ + \omega^-) \int_0^T h_{0i} n^i \, dt = \frac{(\omega^+ + \omega^-)}{c} \oint h_{0i} \, dl^i$$

$$\equiv \frac{(\omega^+ + \omega^-)}{c} \oint \mathbf{h} \cdot \mathbf{dl} \simeq \frac{4\pi}{\lambda} \mathbf{A} \cdot (\nabla \times \mathbf{h}) \simeq 8\pi \frac{\mathbf{\Omega} \cdot \mathbf{A}}{\lambda c} \qquad (3.3.93)$$

since $\nabla \times \mathbf{h} = 2\mathbf{\Omega}/c$. λ is the mean wavelength of the two laser beams, i.e. in this approximation the Sagnac phase shift $(\Delta S)_T$ is given by the (Euclidean) product of enclosed surface $\mathbf{A}$ and angular velocity of axes at rest in the Sagnac platform w.r.t. (local) inertial axes. The same result is found for a Sagnac platform at rest at the Earth's surface in Fermi coordinates (Section 3.4), where the angular velocity is given by the angular velocity of the platform w.r.t. fixed star oriented axes minus the angular velocity from (3.3.71), i.e.

$$(\Delta S)_T = \frac{4\pi}{\lambda} \mathbf{A} \cdot (\nabla \times \mathbf{h}) \qquad (3.3.94)$$

with

$$\nabla \times \mathbf{h} = 2\frac{\mathbf{\Omega}_T}{c} = \frac{2}{c}\left(\mathbf{\Omega}_\oplus + \frac{c}{2}\nabla \times \mathbf{g} - (\gamma + 1)\mathbf{v} \times \frac{\nabla U}{c^2}\right) , \qquad (3.3.95)$$

where the Thomas precession has been absorbed in the last term ($\mathbf{a} = -\nabla U$). The first term represents the classical expression for the Sagnac phase shift due to the Earth's rotation. The two subsequent terms describe the influence of the Lense-Thirring precession and the geodetic precession (corrected for the Thomas term) on the Sagnac phase shift. If we choose the z-axis of an instantaneous rest frame in the direction of the Earth's rotation axis, the center of the Sagnac platform in the x,z-plane, with

$$\mathbf{x} = R_\oplus \left(\cos\lambda, 0, \sin\lambda\right) ,$$

we find

$$\nabla U = -\frac{GM_\oplus}{R_\oplus^2}(\sin\lambda\,\hat{\mathbf{e}}_z + \cos\lambda\,\hat{\mathbf{e}}_x)$$

and

$$\frac{c}{2}\nabla \times \mathbf{g} = (\gamma + 1)\frac{G}{2c^2 R_\oplus^3}\left(\mathbf{J}_\oplus - 3\frac{(\mathbf{J}_\oplus \cdot \mathbf{x})\mathbf{x}}{R_\oplus^2}\right)$$

$$\simeq \frac{1}{5}(\gamma + 1)\Omega_\oplus \frac{GM_\oplus}{c^2 R_\oplus}\left((1 - 3\sin^2\lambda)\hat{\mathbf{e}}_z - 3\sin\lambda\cos\lambda\hat{\mathbf{e}}_x\right) .$$

With

$$\mathbf{A} = A\hat{\mathbf{n}} \quad ; \quad \hat{\mathbf{n}} = (\cos\theta_x, \cos\theta_y, \cos\theta_z)$$

one finally obtains:

$$(\Delta S)_T \simeq \frac{2\omega}{c}A\left(\Omega_\oplus \cos\theta_z + \tilde{\Omega}_{LT} + \tilde{\Omega}_{GPT}\right) \qquad (3.3.96)$$

with

$$\tilde{\Omega}_{LT} = \frac{2}{5}\left(\frac{(\gamma + 1)}{2}\right)\Omega_\oplus\left(\frac{GM_\oplus}{c^2 R_\oplus}\right)\left(-3\sin\lambda\cos\lambda\cos\theta_x + (1 - 3\sin^2\lambda)\cos\theta_z\right)$$

$$\tilde{\Omega}_{\text{GPT}} = (\gamma + 1)\Omega_\oplus \left(\frac{GM_\oplus}{c^2 R_\oplus}\right) \left(\sin\lambda\cos\lambda\cos\theta_x - \cos^2\lambda\cos\theta_z\right) \ .$$

Since

$$\frac{GM_\oplus}{c^2 R_\oplus} \simeq 6.97 \times 10^{-10}$$

we see that both post-Newtonian contributions are about 10^{-10} times smaller than the classical part of the Sagnac phase shift.

Currently various groups plan or already develop such laser gyros with high accuracy, for example the passive, resonant ring laser gyro developed at *Seiler Research Laboratory* (USAF, Colorado Springs) (Rotge et al. 1985) with an area of $\sim 60\,\text{m}^2$ has an expected sensitivity of $\Delta\Omega/\Omega \sim 10^{-10} \sim 10\mu\text{sec/day}$ for averaging times of a few minutes (VLBI: $\Delta\Omega/\Omega \sim 5\times10^{-10}$; averaging times ~ 1 day). It thus can very well be in the future that monitoring UT1 variations and polar motion may be done by means of local inertial systems instead of analyzing photons originating from cosmic distances.

3.4 Reference Frames and Coordinate Systems

Primarily, a reference frame is given independently of the choice of a coordinate system. However, in many cases it is advantageous not to choose the frame and the coordinate system separately from each other. In fact to each local reference frame there exists a set of canonically induced coordinates (as integral curves to the tetrads), and conversely to each coordinate system a set of induced frames. For that reason frequently in the literature one does not distinguish between reference frame and coordinate system, giving rise to a lot of confusion.

3.4.1 Frame Induced Coordinates

One important class of frame induced coordinates are *normal coordinates*. If $p \in \mathcal{M}$, normal coordinates have the general property that geodesics through p appear as straight lines, i.e. the affine connection vanishes in p:

$$\Gamma^\mu_{\nu\sigma}\big|_p = 0 \ . \tag{3.4.1}$$

Since the covariant derivative of the metric tensor vanishes, i.e.

$$g_{\mu\nu,\lambda} = g_{\mu\sigma}\Gamma^\sigma_{\nu\lambda} + g_{\nu\sigma}\Gamma^\sigma_{\mu\lambda} \ , \tag{3.4.2}$$

also the first derivatives of $g_{\mu\nu}$ vanish in p and the second derivatives are given by

$$g_{\mu\nu,\lambda\rho} = \left[g_{\mu\sigma}\Gamma^\sigma_{\nu\lambda,\rho} + g_{\nu\sigma}\Gamma^\sigma_{\mu\lambda,\rho}\right]_p \ . \tag{3.4.3}$$

If we use

$$\Gamma^\mu{}_{(\nu\sigma,\rho)} = 0 \ , \tag{3.4.4}$$

resulting from $\Gamma^{\mu}_{\nu\sigma}|_p = 0$, we find that

$$\Gamma^{\mu}_{\nu\sigma,\rho}\Big|_p = -\frac{2}{3} R^{\mu}_{\ (\nu\sigma)\rho}\Big|_p \qquad (3.4.5)$$

and the components of the metric tensor in the vicinity of p are given by

$$g_{\mu\nu}(x^{\lambda}) = g_{\mu\nu}|_p + \frac{1}{3} R_{\mu\lambda\rho\sigma}\, x^{\lambda} x^{\rho} + \mathcal{O}((x^i x_i)^{3/2}) \ . \qquad (3.4.6)$$

Obviously, there is a whole class of coordinate systems obeying these conditions; it is called the class of *Riemann normal coordinates*. Those normal coordinates that have a direct physical significance are *Fermi normal coordinates* (Synge 1966, Manasse et al. 1963, Misner et al. 1973, Li et al. 1979). They are chosen such that they optimally approximate the concept of inertial coordinates locally for a single observer in free fall ($a = 0$). More precisely, let $\mathcal{G}$ be the geodetic world-line of a single observer, then in Fermi normal coordinates:

$$\begin{array}{lll} g_{\mu\nu}|_{\mathcal{G}} = \eta_{\mu\nu} & (F1) & (3.4.7) \\ \Gamma^{\mu}_{\nu\sigma}|_{\mathcal{G}} = 0 & (F2) \ . & (3.4.8) \end{array}$$

Since we have already discussed the concept of Fermi frames the construction of Fermi normal coordinates with properties (F1) and (F2) is quite simple. Let $\{\mathcal{G}, e_{(\alpha)}\}$ be a Fermi frame of an individual observer $\mathcal{O}$. The Fermi time coordinate t_F then should be given by the proper time τ of $\mathcal{O}$ along $\mathcal{G}$. Finally, let q be in the rest space of $\mathcal{O}$ in $p \in \mathcal{G}$. If q is sufficiently close to p then there will be precisely one space-like geodesic γ_{pq} through p and q with tangent $x^{(i)} e_{(i)}$ in p. If

$$\left((x^{(1)})^2 + (x^{(2)})^2 + (x^{(3)})^2 \right)^{1/2} = |\Delta s| \ , \qquad (3.4.9)$$

where $|\Delta s|$ denotes the proper distance between p and q, then the $x^{(i)}$ represent the spatial Fermi normal coordinates (FNC) of q w.r.t. p. The form of the metric in FNC can again be inferred from (3.4.3), which now should be valid along $\mathcal{G}$. From (F1) and (F2) one finds

$$\Gamma^{\mu}_{\nu\sigma,0} = 0 \qquad (3.4.10)$$

$$\Gamma^{\mu}_{0\nu,\lambda} = R^{\mu}_{\ \nu\lambda 0} \ , \qquad (3.4.11)$$

since $R^{\mu}_{\ \nu\lambda\sigma} = \Gamma^{\mu}_{\sigma\nu,\lambda} - \Gamma^{\mu}_{\lambda\nu,\sigma}$. A third condition can be obtained from the Jacobi equation (Manasse et al. 1963)

$$\Gamma^{\mu}_{ij,k} = -\frac{1}{3}(R^{\mu}_{\ ijk} + R^{\mu}_{\ jik}) \ . \qquad (3.4.12)$$

If the last three relations are inserted into (3.4.3) one obtains

$$g_{00,ij} = 2g_{00}\,\Gamma^0_{0i,j} = -2R_{0i0j} \tag{3.4.13a}$$

$$g_{0m,ij} = -\frac{2}{3}(R_{0imj} + R_{0jmi}) \tag{3.4.13b}$$

$$g_{lm,ij} = -\frac{1}{3}(R_{iljm} + R_{imjl}) \tag{3.4.13c}$$

and

$$g_{00} = -1 - R_{0l0m}|_{\mathcal{G}}\,x^l x^m + \mathcal{O}((x^i x_i)^{3/2}) \tag{3.4.14a}$$

$$g_{0i} = -\frac{2}{3}R_{0lim}|_{\mathcal{G}}\,x^l x^m + \mathcal{O}((x^i x_i)^{3/2}) \tag{3.4.14b}$$

$$g_{ij} = \delta_{ij} - \frac{1}{3}R_{iljm}|_{\mathcal{G}}\,x^l x^m + \mathcal{O}((x^i x_i)^{3/2})\ . \tag{3.4.14c}$$

Of great practical importance are geocentric FNC, in which the influence of Sun, Moon and planets manifests itself only in second order in x^i as tidal forces, where the gravitational field of the Earth, however, has to be additionally introduced into the metric (3.4.14) (Ashby et al. 1984, Mashhoon 1985). To this end one can consider the metric (3.4.14) as fixed background metric and then solve the Einstein field equations for the outer field of the Earth; one finds that the nonlinear interaction terms of form $U_\oplus R_{\mu\nu\sigma\lambda}$ are negligibly small and therefore, to post-Newtonian order, the curvature terms from (3.4.14) can simply be added to the metric (A1.4) for the isolated Earth.

If the world-line of the observer is not a geodesic one no longer can require that all Christoffel symbols vanish along $\mathcal{G}$ since the acceleration $\mathbf{a}$ experienced by a gravimeter is just determined by these quantities. With respect to a Fermi tetrad one finds (e.g. Misner et al. 1973):

$$\Gamma^{(i)}_{\ (0)(0)} = \Gamma^{(0)}_{\ (0)(i)} = a^{(i)} \tag{3.4.15}$$

and for the metric in corresponding normal coordinates one finds (3.4.14), the -1 in g_{00}, however, being replaced by

$$-\left(1 + \frac{\mathbf{a}\cdot\mathbf{x}}{c^2}\right)^2\ .$$

Arbitrary spatial axes $e_{(i)}$ differ from Fermi axes by a spatial rotation with

$$\omega_{(i)(j)} = (D_{\mathbf{u}}e_{(i)}, e_{(j)})\ ,$$

leading to

$$\Gamma^{(j)}_{\ (k)(0)} = -\epsilon_{(0)(i)(j)(k)}\,\Omega^{(i)}\ . \tag{3.4.16}$$

In the local *proper reference frame* of the observer for small angular velocities the metric is given to a good approximation by:

$$g_{00} = -\left(1 + \frac{\mathbf{a} \cdot \mathbf{x}}{c^2}\right)^2 - R_{(0)(i)(0)(j)}|_{\mathcal{G}} \, x^{(i)} x^{(j)} \qquad (3.4.17a)$$

$$g_{0i} = -\epsilon_{(i)(j)(k)} \, \Omega^{(j)} x^{(k)} - \frac{2}{3} R_{(0)(j)(i)(k)}|_{\mathcal{G}} \, x^{(j)} x^{(k)} \qquad (3.4.17b)$$

$$g_{ij} = \delta_{ij} - \frac{1}{3} R_{(i)(k)(j)(l)}|_{\mathcal{G}} \, x^{(k)} x^{(l)} \; . \qquad (3.4.17c)$$

3.4.2 Coordinate Induced Frames

If a (global) coordinate system $x^\mu = (t, \mathbf{x})$ is given, then a platform is said to be at rest w.r.t. the coordinates if the observers γ are given by $\mathbf{x} = \text{const.}$ lines. In this case the 4-velocity $\mathbf{u}$ has only a time-like component, i.e.

$$\mathbf{u} = \mathbf{e}_{(0)} = \frac{1}{\sqrt{|g_{00}|}} \frac{\partial}{\partial t} \quad ; \quad u^\mu = e^\mu_{(0)} = (\frac{1}{\sqrt{|g_{00}|}}, 0, 0, 0) \; . \qquad (3.4.18)$$

For the spatial triad of the corresponding reference frame it follows that:

$$e_{(m)0} = 0 \qquad (3.4.19)$$

and the normalization of $e_{(m)}$ requires that

$$g^{ij} e_{(m)i} e_{(n)j} = \delta_{mn}$$

or

$$W \hat{g}^{-1} W^T = 1 \qquad (3.4.20)$$

with

$$W_{mi} = e_{(m)i} \quad \text{and} \quad \hat{g}^{-1} = (g^{ij}) \; .$$

If we write this in the form

$$DD^T = 1$$

with $D = W \hat{g}^{-1/2}$, we see that $D \in SO(3)$. This simply expresses the fact that the spatial triad in the rest frame of the observer without additional constraint is determined only up to a spatial rotation. The condition $D = 1$, however, selects that spatial triad that in some sense optimally approximates the tangent unit vectors to the coordinate lines.

Definition. *A tetrad field is called induced by coordinates x^μ, if*
  i) $e_{(0)} = |g_{00}|^{-1/2} \partial/\partial t$
  ii) $e_{(m)i} = (\hat{g}^{1/2})_{mi}$

Since $\hat{g}$ is positive definite $\hat{g}^{1/2}$ always exists and hence also the coordinate induced (natural) tetrad field. If the metric field $g_{\mu\nu}$ is diagonal the induced tetrad is simply given by $e_{(\alpha)} = |g_{\alpha\alpha}|^{-1/2} \partial/\partial x^\alpha$.

3.5 Station Coordinates and the Problem of Length

The *determination of "coordinates"* of selected points (stations etc.) at the
Earth's surface is one of the major concerns of geodesy and one can raise
the question of how such coordinates can be chosen in a relativistic context.
According to the last Chapters it is obvious that each choice of coordinates
necessarily involves conventions and/or constructions, e.g. out of a frame, if we
leave the ground of Newtonian space-time theory. If we work in a terrestrial
system it suggests employing a corresponding system of Fermi normal coordinates. Now, for most applications we can neglect the precession of Fermi axes
w.r.t. fixed star oriented axes (by $2''$ per century). In Section 3.3.3 we have
shown how such fixed star oriented axes $e'_{(\alpha)}$ co-moving with the Earth can be
constructed in barycentric PPN coordinates: starting from the standard PPN
coordinate induced tetrad (at rest w.r.t. the coordinates)

$$e_{(t)} = \eta_+ \frac{\partial}{\partial t} \quad ; \quad e_{(i)} = \eta_- \gamma \frac{\partial}{\partial x^i} \tag{3.5.1}$$

with

$$\eta_\epsilon = 1 + \epsilon \frac{U(\mathbf{x}_\oplus)}{c^2} \tag{3.5.2}$$

they can be obtained by means of a (local) Lorentz transformation Λ, i.e.

$$e'_{(\alpha)} = \Lambda_{(\alpha)}{}^{(\beta)} e_{(\beta)} \tag{3.5.3}$$

with ($\mathbf{v} = \mathbf{v}_\oplus$):

$$\Lambda_{(0)}{}^{(0)} = \gamma' \simeq 1 + \frac{1}{2}\frac{\mathbf{v}^2}{c^2} \tag{3.5.4a}$$

$$(\Lambda_{(i)}{}^{(0)}) = (\Lambda_{(0)}{}^{(i)}) = \gamma'\frac{\mathbf{v}}{c} \simeq \frac{\mathbf{v}}{c} \tag{3.5.4b}$$

$$(\Lambda_{(i)}{}^{(j)}) = 1 + \frac{(\gamma'-1)}{\mathbf{v}^2}\mathbf{v} \otimes \mathbf{v} \simeq 1 + \frac{1}{2}\frac{\mathbf{v} \otimes \mathbf{v}}{c^2} \quad . \tag{3.5.4c}$$

To our accuracy it was sufficient to take the coordinate velocity of the
Earth in the Lorentz boost. Suppose such geocentric coordinates, for example
of a station at the Earth's surface, are given and we ask about the corresponding
barycentric PPN coordinates. This problem is of great practical significance
since, for example by satellite laser ranging geocentric "station vectors" are
determined whereas the ephemerides programs used for the computation of
the tidal forces are usually written in barycentric coordinates. To this end
we want to consider the Earth as a "quasi-rigid" body after the deformations
or crustal motions (tectonic plate motions, Earth- and ocean-tides etc.) have
been treated in a "Newtonian" manner. It is useful to study this problem
first without the solar gravitational field, i.e. as a problem within the frame of

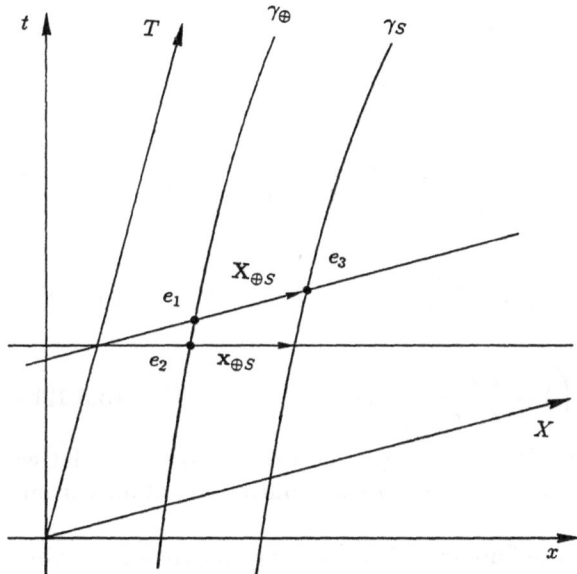

Fig. 3.6. Geometry in the problem of station coordinates

special relativity. According to our discussion in Section 3.1 geocentric "station vectors" will be defined in a nonrotating, instantaneous system. Let $(t, \mathbf{x})$ denote the barycentric coordinate system, $(T, \mathbf{X})$ the instantaneous geocentric coordinates, then both are related by the Lorentz transformation ($\mathbf{v} = \mathbf{v}_\oplus$)

$$\Delta T = \gamma'(\Delta t - \mathbf{v} \cdot \Delta \mathbf{x}/c^2) \simeq \Delta t - \mathbf{v} \cdot \Delta \mathbf{x}/c^2 \qquad (3.5.5a)$$

$$\Delta \mathbf{X} = \Delta \mathbf{x} + \frac{(\gamma' - 1)}{\mathbf{v}^2}(\mathbf{v} \cdot \Delta \mathbf{x})\mathbf{v} - \gamma' \mathbf{v} \Delta t \simeq \Delta \mathbf{x} - \mathbf{v} \Delta t \ . \qquad (3.5.5b)$$

Let S denote a station and let (Fig. 3.6)

$$\mathbf{x}_{\oplus S}(t) = \mathbf{x}_S(t) - \mathbf{x}_\oplus(t) \ . \qquad (3.5.6)$$

Let us now focus upon the two points (events) e_1 and e_3 from Fig. 3.6, that should define the geocentric station vector $\mathbf{X}_{\oplus S}$. Now, $\mathbf{X}_{\oplus S}$ should lie in a surface of geocentric simultaneity, i.e.

$$\mathbf{X}_{\oplus S} = \mathbf{X}_S(T_E) - \mathbf{X}_\oplus(T_E) \qquad (3.5.7)$$

with $\Delta T(e_1, e_3) = 0$. From this one obtains acording to (3.5.5a):

$$\Delta t \simeq \mathbf{v} \cdot \Delta \mathbf{x}/c^2 \ . \qquad (3.5.8)$$

With

$$\mathbf{b}_{\oplus S} \equiv \mathbf{x}_S(t_S) - \mathbf{x}_\oplus(t_E) \qquad (3.5.9)$$

it follows that

$$\mathbf{X}_{\oplus S} \simeq \left(1 - \frac{1}{2}\frac{\mathbf{v} \otimes \mathbf{v}}{c^2}\right)\mathbf{b}_{\oplus S} . \tag{3.5.10}$$

We finally want to express $\mathbf{b}_{\oplus S}$ by $\mathbf{x}_{\oplus S}$. Since ($\mathbf{v}_S \simeq \mathbf{v}_\oplus = \mathbf{v}$):

$$\mathbf{b}_{\oplus S} \simeq \mathbf{x}_S(t_E) + \frac{1}{c^2}(\mathbf{x}_{\oplus S} \cdot \mathbf{v})\mathbf{v} - \mathbf{x}_\oplus(t_E) \simeq \left(1 + \frac{\mathbf{v} \otimes \mathbf{v}}{c^2}\right)\mathbf{x}_{\oplus S} \tag{3.5.11}$$

and we simply get

$$\mathbf{X}_{\oplus S} \simeq \left(1 + \frac{1}{2}\frac{\mathbf{v} \otimes \mathbf{v}}{c^2}\right)\mathbf{x}_{\oplus S} \tag{3.5.12a}$$

or

$$\mathbf{x}_{\oplus S} \simeq \left(1 - \frac{1}{2}\frac{\mathbf{v} \otimes \mathbf{v}}{c^2}\right)\mathbf{X}_{\oplus S} . \tag{3.5.12b}$$

This essentially diurnal variation of the barycentric station vector by a relative amount of $\sim 10^{-8}$ represents nothing but the Lorentz contraction of measuring rods in the velocity direction.

If gravitational effects from the Sun are taken into account then according to the spatial metric $g_{ij} = (1 + 2\gamma U/c^2)\delta_{ij}$ one would expect that the last expression is replaced by

$$\mathbf{x}_{\oplus S} \simeq \eta_{-\gamma}\left(1 - \frac{1}{2}\frac{\mathbf{v} \otimes \mathbf{v}}{c^2}\right)\mathbf{X}_{\oplus S} . \tag{3.5.13}$$

We now want to derive such a relation by means of frame induced co-ordinates. We have already approximately determined the tetrad $e'_{(\alpha)}$ co-moving with the Earth in (3.5.3) and we can now ask about the coordinates $X^\mu = (T, \mathbf{X})$ in which $e'_{(\alpha)}$ is given by $e'_{(\alpha)} = \partial/\partial X^\mu$. One finds that these are determined by (Fukushima et al. 1986, Fujimoto et al. 1986)

$$\Delta X^{(\alpha)} = \left[e'^\mu_{(\alpha)}\right]^{-1} \Delta x^\mu \tag{3.5.14a}$$

or

$$\Delta x^\mu = \left[e'^\mu_{(\alpha)}\right] \Delta X^{(\alpha)}, \tag{3.5.14b}$$

where the *Frobenius matrix* $[e'^\mu_{(\alpha)}]$ according to (3.5.3) is given by†

$$\left[e'^\mu_{(\alpha)}\right] = \begin{bmatrix} \gamma'\eta_+ & \mathbf{v}/c \\ \mathbf{v}/c & \eta_{-\gamma}(1 + \mathbf{v} \otimes \mathbf{v}/c^2) \end{bmatrix} + \mathcal{O}(\epsilon^3) \tag{3.5.15a}$$

---

† In some sense relation (3.5.14) represents the generalization of the "global Lorentz transformation" in (3.5.5). Such a characterization, however, must be handled with care: in a curved space-time the Lorentz transformation operates only *locally*, i.e. it transforms only (basis) 4-vectors at a point $p \in \mathcal{M}$. In Minkowski-space we can look upon global inertial coordinates as being induced by inertial axes that can be related by global Lorentz transformations.

or

$$\left[e'^{\mu}_{(\alpha)}\right]^{-1} = \begin{bmatrix} \gamma'\eta_- & -\mathbf{v}/c \\ -\mathbf{v}/c & \eta_{+\gamma}(1 + \mathbf{v} \otimes \mathbf{v}/c^2) \end{bmatrix} + \mathcal{O}(\epsilon^3) \ . \tag{3.5.15b}$$

We would now like to address the problem of units (e.g. Fukushima et al. 1986), before finishing the discussion of station coordinates. The General Conference of Mass and Weights defines the international unit of the SI-meter as (17th CGPM, 17th October 1983):

"Le mètre est la longueur du trajet parcouru dans le vide par la lumière pendant une durée de 1/299 792 458 de seconde."

that after translation reads

"The meter is the length of the path travelled by light in vacuum during a time interval of 1/299 792 458 of a second."

Just as in the definition of the SI second the unit of the SI meter is given via some measured quantity that leads to the difficulties discussed in Section 3.2. In analogy to the time problem one might tend to renormalize the barycentric SI unit of length. Remember that in the time problem we switched from the barycentric coordinate time t to TDB with:

$$\Delta t' \equiv \Delta(TDB) = (1 - \langle U + \mathbf{v}^2/2 \rangle /c^2|_{\text{geoid}}) \Delta t \equiv \alpha_G \Delta t. \tag{3.5.16}$$

Such a renormalization of the unit of length in the barycentric system can be performed in a variety of ways. E.g. we can require the numerical constancy of c, i.e. we can introduce a barycentric unit of length that differs from the barycentric SI meter by a factor of α_G, i.e.

$$\mathbf{x}' = \alpha_G \cdot \mathbf{x} \ . \tag{3.5.17}$$

In this *new* unit of length the equations (3.5.14a,b) read:

$$\Delta X^{(\alpha)} = [\alpha_G \cdot e'^{\mu}_{(\alpha)}]^{-1} \Delta x'^{\mu} \tag{3.5.18a}$$

$$\Delta x'^{\mu} = [\alpha_G \cdot e'^{\mu}_{(\alpha)}] \Delta X^{(\alpha)} \tag{3.5.18b}$$

or explicitly (Fukushima et al. 1986 (3-21), (3-22)) $(\beta_G = \alpha_G^{-1})$:

$$\Delta T = \beta_G \gamma'\eta_- \Delta t' - \mathbf{v} \cdot \mathbf{x}'/c^2 \tag{3.5.19a}$$

$$\Delta X = \beta_G \eta_{+\gamma} \left(1 + \frac{1}{2}\frac{\mathbf{v} \otimes \mathbf{v}}{c^2}\right) \Delta x' - \mathbf{v} \cdot \Delta t' \tag{3.5.19b}$$

and

$$\Delta t' = \alpha_G \gamma' \eta_+ \Delta T + \mathbf{v} \cdot \Delta \mathbf{X}/c^2 \qquad (3.5.20a)$$

$$\Delta \mathbf{x}' = \alpha_G \eta_{-\gamma} \left(1 + \frac{1}{2}\frac{\mathbf{v} \otimes \mathbf{v}}{c^2}\right) \Delta \mathbf{X} + \mathbf{v} \cdot \Delta T \; . \qquad (3.5.20b)$$

Note that the factor in front of ΔT in (3.5.20a) is just given by

$$\alpha_G \gamma' \eta_+ = 1 + \frac{1}{c^2}\left(U + \frac{1}{2}\mathbf{v}\right) - \frac{1}{c^2}\left\langle U + \frac{1}{2}\mathbf{v}\right\rangle = 1 + \left(U + \frac{1}{2}\mathbf{v}^2\right)_p \; . \quad (3.5.21)$$

We can now repeat the argument for the field free case with the transformation (3.5.5) being replaced by (3.5.19). One obtains ($\mathbf{v} = \mathbf{v}_\oplus; \kappa_G = \langle U + \mathbf{v}^2/2\rangle/c^2$):

$$\mathbf{X}_{\oplus S} = \beta_G \eta_{+\gamma}\left(1 + \frac{1}{2}\frac{\mathbf{v} \otimes \mathbf{v}}{c^2}\right)\mathbf{x}'_{\oplus S} = \left[1 + \left(\gamma\frac{U}{c^2} + \frac{1}{2}\frac{\mathbf{v} \otimes \mathbf{v}}{c^2} + \kappa_G\right)\right]\mathbf{x}'_{\oplus S}$$

$$\mathbf{x}'_{\oplus S} = \left[1 - \left(\gamma\frac{U}{c^2} + \frac{1}{2}\frac{\mathbf{v} \otimes \mathbf{v}}{c^2} + \kappa_G\right)\right]\mathbf{X}_{\oplus S} \; . \qquad (3.5.22)$$

If the barycentric SI-units are artificially rescaled then this has the consequence that all dimensional quantities have to be rescaled in the same manner: e.g. barycentric (GM) values differ from corresponding geocentric values by a factor of α_G. For that reason the usefulness of the IAU recommendation for the rescaling of barycentric units seems to be very doubtful.

4. Celestial Mechanics

4.1 Post-Newtonian Motion of Point Masses†

Like in all metric theories of gravity uncharged test bodies move on geodesics
in the PPN formalism. Hence, for test bodies the equation of motion is given
by

$$\frac{d^2 x^\mu}{d\lambda^2} + \Gamma^\mu_{\nu\sigma} \frac{dx^\mu}{d\lambda} \frac{dx^\sigma}{d\lambda} = 0 \tag{4.1.1}$$

if λ is an affine parameter. Instead of an affine parameter one frequently uses
the coordinate time t and obtains the coordinate acceleration of a test body
from (Weinberg 1972):

$$\frac{d^2 x^i}{dt^2} = \left(\frac{dt}{d\lambda}\right)^{-1} \frac{d}{d\lambda} \left[\left(\frac{dt}{d\lambda}\right)^{-1} \frac{dx^i}{d\lambda}\right] = \left(\frac{dt}{d\lambda}\right)^{-2} \frac{d^2 x^i}{d\lambda^2} - \left(\frac{dt}{d\lambda}\right)^{-3} \frac{d^2 t}{d\lambda^2} \frac{dx^i}{d\lambda}$$

$$= -\Gamma^i_{\nu\lambda} \frac{dx^\nu}{dt} \frac{dx^\lambda}{dt} + \Gamma^0_{\nu\lambda} \frac{dx^\nu}{dt} \frac{dx^\lambda}{dt} \frac{dx^i}{dt}$$

$$= -\Gamma^i_{00} - 2\Gamma^i_{0j} v^j - \Gamma^i_{jk} v^j v^k + \left(\Gamma^0_{00} + 2\Gamma^0_{0j} v^j + \Gamma^0_{jk} v^j v^k\right) v^i \ . \tag{4.1.2}$$

Substituting the Christoffel symbols from (A1.2) we obtain:

$$\frac{d\mathbf{v}}{dt} = \nabla(U - (\beta + \gamma)U^2 + \Phi) - \frac{(\gamma+1)}{2} \frac{\partial \boldsymbol{\zeta}}{\partial t} - \frac{1}{2} \frac{\partial^2}{\partial t^2} \nabla \chi + \frac{(\gamma+1)}{2} \mathbf{v} \times (\nabla \times \boldsymbol{\zeta})$$

$$- (2\gamma+1)\mathbf{v} \frac{\partial U}{\partial t} - 2(\gamma+1)\mathbf{v}(\mathbf{v} \cdot \nabla)U + \gamma v^2 \nabla U \tag{4.1.3}$$

with

$$\chi(t, \mathbf{x}) \equiv -G \int \rho' |\mathbf{x} - \mathbf{x}'| d^3 x' \tag{4.1.4}$$

and (A1.1)

$$\zeta_i \equiv -4V_i \ . \tag{4.1.5}$$

---

† Metric, Christoffel symbols etc. for a system of point masses can be found in A.1 of the
Appendix.

The potential χ has the properties

$$\chi_{,0i} = V_i - W_i \quad ; \quad U = -\frac{1}{2}\Delta\chi \tag{4.1.6}$$

and we used Γ^i_{00} in the form:

$$\Gamma^i_{00} = -U_{,i} + \frac{\partial}{\partial x^i}\left((\beta + \gamma)U^2 - \Phi\right) + \frac{(\gamma + 1)}{2}\frac{\partial}{\partial t}\zeta_i + \frac{1}{2}\frac{\partial^3\chi}{\partial t^2\partial x^i} \; . \tag{4.1.7}$$

For an ensemble of N "point masses", we can obtain the coordinate acceleration of the ath particle of vanishing mass ($m_a \to 0$) according to (4.1.2) by the following substitutions in the Christoffel symbols (A1.15):

$$\mathbf{x} \to \mathbf{x}_a \quad ; \quad r_a \to r_{ab}$$
$$a \to b \quad ; \quad b \to c$$
$$\sum_a \to \sum_{b \neq a} \quad ; \quad \sum_{b \neq a} \to \sum_{c \neq a,b}$$

etc. However, we would like to derive such an equation of motion by means of a Lagrangian which has the advantage that we can easily cover also the case with $m_a \neq 0$.

The geodetic equation (4.1.1) can be obtained by means of the variational principle

$$\delta\int d\tau = \delta\int\left(\frac{d\tau}{dt}\right)dt = 0 \tag{4.1.8}$$

with $d\tau^2 = -ds^2$. Now,

$$\left(\frac{d\tau}{dt}\right)^2 = -g_{\mu\nu}\left(\frac{dx^\mu}{dt}\right)\left(\frac{dx^\nu}{dt}\right) = 1 - \mathbf{v}^2 - \overset{(2)}{g}_{00} - \overset{(4)}{g}_{00} - 2\overset{(3)}{g}_{0i}v^i - \overset{(2)}{g}_{ij}v^iv^j, \tag{4.1.9}$$

if the superscript number in brackets indicates the PN order. The geodetic equation is then equivalent to the Euler-Lagrange equation with the Lagrangian:

$$L_a = 1 - \frac{d\tau}{dt}$$

$$= \frac{1}{2}\mathbf{v}_a^2 + \frac{1}{8}\mathbf{v}_a^4 + \frac{1}{2}(\overset{(2)}{g}_{00} + \overset{(4)}{g}_{00}) + \overset{(3)}{g}_{0i}v_a^i$$

$$+ \frac{1}{2}\overset{(2)}{g}_{ij}v_a^iv_a^j + \frac{1}{4}\mathbf{v}_a^2\overset{(2)}{g}_{00} + \frac{1}{8}(\overset{(2)}{g}_{00})^2 \; . \tag{4.1.10}$$

Inserting the metric from (A1.14) yields, after renaming the indices,

$$L_a = \frac{1}{2}\mathbf{v}_a^2 + \frac{1}{8}\mathbf{v}_a^4 +$$
$$+ \frac{1}{2}\sum_{b\neq a}\frac{m_b}{r_{ab}}\Big[2 + (2\gamma+1)(\mathbf{v}_a^2 + \mathbf{v}_b^2)$$
$$- (4\gamma+3)\mathbf{v}_a\cdot\mathbf{v}_b - (\mathbf{v}_a\cdot\mathbf{n}_{ab})(\mathbf{v}_b\cdot\mathbf{n}_{ab})\Big]$$
$$- \frac{(2\beta-1)}{2}\sum_{b,c\neq a}\frac{m_b\,m_c}{r_{ab}\,r_{ac}} - (2\beta-1)\sum_{b\neq a}\sum_{c\neq a,b}\frac{m_b\,m_c}{r_{ab}\,r_{bc}} \qquad (4.1.11)$$

where

$$\mathbf{n}_{ab} = \mathbf{x}_{ab}/r_{ab} \quad ; \quad \mathbf{x}_{ab} = \mathbf{x}_b - \mathbf{x}_a \ .$$

This Lagrange function can easily be generalized onto a system of N particles moving in a self-consistent field: $\mathcal{L}_N$ has to be a symmetrical expression in $(m_a, \mathbf{x}_a, \mathbf{v}_a; a = 1, \ldots, N)$ and possessing the property

$$\lim_{m_a\to 0}\frac{\partial}{\partial m_a}\mathcal{L}_N = L_a \ . \qquad (4.1.12)$$

This leads us to the PPN Lagrange function

$$\mathcal{L}_N = \sum_a m_a\left(\frac{1}{2}\mathbf{v}_a^2 + \frac{1}{8}\mathbf{v}_a^4\right)$$
$$+ \frac{1}{2}\sum_{\substack{a,b\\b\neq a}}\frac{m_a m_b}{r_{ab}}\Big[1 + (2\gamma+1)\mathbf{v}_a^2$$
$$- \frac{1}{2}(4\gamma+3)\mathbf{v}_a\cdot\mathbf{v}_b - \frac{1}{2}(\mathbf{v}_a\cdot\mathbf{n}_{ab})(\mathbf{v}_b\cdot\mathbf{n}_{ab})\Big]$$
$$- \frac{(2\beta-1)}{2}\sum_{\substack{a,b\\b\neq a}}\sum_{c\neq a}\frac{m_a m_b\,m_c}{r_{ab}\,r_{ac}} \ . \qquad (4.1.13)$$

The corresponding Euler-Lagrange equations of motion read:

$$\frac{d\mathbf{v}_a}{dt} = \sum_{b\neq a}\frac{m_b\mathbf{x}_{ab}}{r_{ab}^3}\Big[1 - 2(\beta+\gamma)\sum_{c\neq a}\frac{m_c}{r_{ac}}$$
$$+ \sum_{c\neq a,b}m_c\Big(-\frac{(2\beta-1)}{r_{bc}} + \frac{\mathbf{x}_{ab}\cdot\mathbf{x}_{bc}}{2r_{bc}^3}\Big) - (2\gamma+2\beta+1)\frac{m_a}{r_{ab}}$$
$$+ \gamma\mathbf{v}_a^2 - (2\gamma+2)\mathbf{v}_a\cdot\mathbf{v}_b + (\gamma+1)\mathbf{v}_b^2 - \frac{3}{2}\Big(\frac{\mathbf{v}_b\cdot\mathbf{x}_{ab}}{r_{ab}}\Big)^2\Big] +$$

$$+ \sum_{b \neq a} m_b \left[\frac{\mathbf{x}_{ab}}{r_{ab}^3} \cdot ((2\gamma + 2)\mathbf{v}_a - (2\gamma + 1)\mathbf{v}_b) \right] (\mathbf{v}_b - \mathbf{v}_a)$$

$$+ \frac{(4\gamma + 3)}{2} \sum_{b \neq a} \sum_{c \neq a, b} \frac{m_b}{r_{ab}} \frac{m_c \mathbf{x}_{bc}}{r_{bc}^3} \, , \tag{4.1.14}$$

that are also called (PPN) *Einstein-Infeld-Hoffmann* (EIH) equations of motion (Einstein et al. 1938, see also Lorentz et al. 1917). For the case of a test body with $m_a \to 0$, moving along a geodesic, this result can directly be obtained from (4.1.2) by the indicated substitutions in the Christoffel symbols.

4.2 Motion of Satellites

Before we start with the main discussion of "relativistic effects" in the motion of artificial satellites we first would like to say what we understand by "relativistic effects". In the post-Newtonian (any relativistic) formalism the expressions for the observables depend upon the *reference frame* of the observer (relativity) and non-observables like positions and velocities of satellites, planets etc. depend also upon the choice of the *coordinate system*. Even after having chosen the reference frame and the coordinates to answer the question: "how large are the relativistic effects?" one faces in addition the following difficulty: if we simply add the corresponding "post-Newtonian correction terms" to a Newtonian theory with well defined initial conditions e.g. for positions and velocities of moving bodies (satellites, planets etc.) and dynamical variables (mass multipole moments etc.) then the solutions will very rapidly show large deviations as a function of time. This is because each theory possesses its own set of initial conditions and dynamical variables that have to be fitted to observational data ("the Newtonian mass of the Earth $\neq$ post-Newtonian mass of the Earth" etc.). Strictly speaking, it only makes sense to comprehend the residuals occuring in a comparison between theoretically predicted observables and measured data as the "influence of relativistic effects" if initial conditions and dynamical variables are optimally fitted to the data in both cases. However, in practice this procedure of parameter estimation from measured time series is very cumbersome and, therefore, one often has to be content with a comparison for equal initial conditions and dynamical parameters. If one is not too cautious one might also like to talk about relativistic effects in non-observables. In that case one has to be fully aware of the fact that the magnitude of such "relativistic effects" depends not only upon the reference frame of the observer but also upon the *arbitrary choice* of the coordinates. We would like to discuss this point in more detail for the example of the motion of artificial satellites.

4.2.1 Spherical Field of the Earth and the Problem of Coordinates

Since the Schwarzschild radius of the Earth ($2GM_\oplus/c^2$) is about 1 cm one would expect this length scale to determine the "magnitude of relativistic effects" in the vicinity of the Earth and in a geocentric reference frame. On the other hand we have seen in Section 3.5 that from a barycentric point of view all "position vectors" (e.g. that of LAGEOS) moving with the Earth due to the Lorentz contraction vary periodically with a relative amount of about 10^{-8} (i.e. the semi-major axis of the LAGEOS orbit with $a \simeq 12\,270$ km thereby varies by about 10 cm!). Since we have already discussed the relationship between geocentric and barycentric station vectors (including satellite position vectors etc.) in this Section we will take the geocentric standpoint.

In standard PN coordinates $(t, \mathbf{x})$ the equation of motion for a satellite in the spherical gravitational field reads (4.1.3; A2.6)

$$\frac{d\mathbf{v}}{dt} = -\frac{\mu\mathbf{x}}{r^3} + \frac{\mu}{c^2}\left[2(\beta + \gamma)\frac{\mu}{r^4}\mathbf{x} - \gamma\frac{v^2}{r^3}\mathbf{x} + 2(\gamma+1)\frac{(\mathbf{x}\cdot\mathbf{v})}{r^3}\mathbf{v}\right] \qquad (4.2.1)$$

with

$$\mu \equiv GM_\oplus \equiv GM \ .$$

If we now choose a new radial coordinate r' with[†]

$$r' = r + \alpha\frac{\mu}{c^2} \ , \qquad (4.2.2)$$

then the satellite acceleration is given by

$$\frac{d\mathbf{v}}{dt} = -\frac{\mu\mathbf{x}}{r^3}$$
$$+ \frac{\mu}{c^2}\left[2(\beta + \gamma - \alpha)\frac{\mu}{r^4}\mathbf{x} - (\gamma + \alpha)\frac{v^2}{r^3}\mathbf{x}\right.$$
$$\left. + 3\alpha\frac{(\mathbf{x}\cdot\mathbf{v})^2}{r^5}\mathbf{x} + 2(\gamma + 1 - \alpha)\frac{(\mathbf{x}\cdot\mathbf{v})}{r^3}\mathbf{v}\right] \ , \qquad (4.2.3)$$

where we have omitted the primes for $r, \mathbf{x}$ and $\mathbf{v}$. In Brumberg (1972) the motion of artificial satellites is generally discussed with the gauge parameter α. Here, we would like to restrict the discussion of the coordinate problem to an inspection of the Keplerian "osculating elements".[‡]

[†] The constant α is the gauge parameter introduced by Brumberg (1972).

[‡] Solutions of (4.2.3) for $\alpha = 0$ (standard PN coordinates) can be found in A.2, where the full post-Newtonian two-body problem is treated.

Let us write (4.2.1) in the form

$$\frac{d\mathbf{v}}{dt} = -\frac{\mu \mathbf{x}}{r^3} + \mathbf{a}_{\text{PN}} \, , \tag{4.2.4}$$

i.e. let us formally interpret (4.2.1) as the equation of motion of a perturbed Keplerian problem in our chosen coordinate system that we can treat with the means of classical perturbation theory. We can interpret $\mathbf{a}_{\text{PN}}$ as perturbing function and decompose it in the usual Euclidean sense into radial, perpendicular and normal component according to:

$$\mathbf{a} = S\mathbf{n} + T\mathbf{m} + W\mathbf{k} \tag{4.2.5}$$

with

$$\mathbf{n} = \begin{pmatrix} \cos u \cos \Omega - \sin u \sin \Omega \cos I \\ \cos u \sin \Omega + \sin u \cos \Omega \cos I \\ \sin u \sin I \end{pmatrix} \tag{4.2.6a}$$

and

$$\mathbf{m} = \frac{\partial}{\partial u}\mathbf{n} \quad ; \quad \mathbf{k} = \frac{1}{\sin u}\left(\frac{\partial}{\partial I}\mathbf{n}\right) \, . \tag{4.2.6b}$$

Here, $u = \omega + f$ denotes the angle between the *line of nodes* (the intersection between the orbital plane and the reference plane, e.g. the equatorial plane of the Earth in the case of satellite orbits) and the position vector, ω denotes the *argument of the perigee* and f the *true anomaly*. I is the inclination of the orbit w.r.t. the reference plane and Ω the *longitude of the ascending node* (the angle between the reference direction, e.g. the direction to the vernal equinox, and the line of nodes, Fig. 4.1).

The perturbation equations (Lagrange's planetary equations) for the orbital elements (a: *semi-major axis*, e: *numerical eccentricity*, M: *mean anomaly*) read (e.g. Roy 1978):

$$\frac{da}{dt} = \frac{2}{n\sqrt{1-e^2}}\left(Se \sin f + T\frac{p}{r}\right) \tag{4.2.7a}$$

$$\frac{de}{dt} = \frac{\sqrt{1-e^2}}{na}\left(S \sin f + T(\cos f + \cos E)\right) \tag{4.2.7b}$$

$$\frac{dI}{dt} = \frac{r \cos(\omega + f)}{na^2\sqrt{1-e^2}}W \tag{4.2.7c}$$

$$\frac{d\omega}{dt} = -\cos I \frac{d\Omega}{dt} + \frac{\sqrt{1-e^2}}{nae}\left[-S \cos f + T\left(1 + \frac{r}{p}\right)\sin f\right] \tag{4.2.7d}$$

$$\frac{d\Omega}{dt} = \frac{r \sin(\omega + f)}{na^2 \sin I\sqrt{1-e^2}}W \tag{4.2.7e}$$

$$\frac{dM}{dt} = n - \sqrt{1-e^2}\left(\frac{d\omega}{dt} + \cos I \frac{d\Omega}{dt}\right) - S\frac{2r}{na^2} \tag{4.2.7f}$$

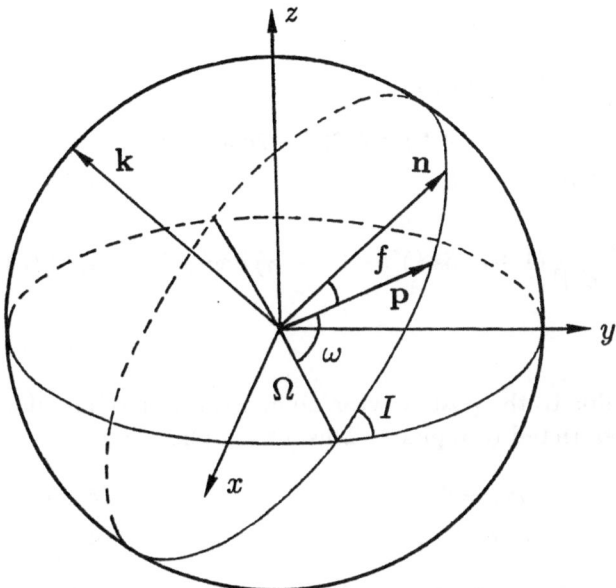

Fig. 4.1. Geometry in the unperturbed Newtonian Kepler problem.

with

$$p = a(1 - e^2) \quad ; \quad n^2 a^3 = \mu \ .$$

Here n denotes the mean motion (mean angular velocity) in the orbit. Instead of the argument of the perigee ω and the mean anomaly M one frequently uses the *longitude of perihelion* ϖ and the *mean longitude at the epoch* ϵ given by

$$\varpi = \omega + \Omega \tag{4.2.8}$$

and

$$M + \varpi = \int n \, dt + \epsilon \ . \tag{4.2.9}$$

These quantities obey the relations:

$$\frac{d\varpi}{dt} = 2\frac{d\Omega}{dt}\sin^2\frac{I}{2} + \frac{\sqrt{1 - e^2}}{nae}\left[-S\cos f + T\left(1 + \frac{r}{p}\right)\sin f\right] \tag{4.2.10a}$$

and

$$\frac{d\epsilon}{dt} = \frac{e^2}{1 + \sqrt{1 - e^2}}\frac{d\varpi}{dt} + 2\frac{d\Omega}{dt}(1 - e^2)^{1/2}\left(\sin^2\frac{I}{2}\right) - \frac{2rS}{na^2} \ . \tag{4.2.10b}$$

The post-Newtonian acceleration (4.2.1) yields the components S, T and W of the perturbing acceleration (cf. A2.77 for $\alpha = 0$, $\nu = 0$) as:

$$S = \frac{\mu}{c^2} \frac{\mu}{a^3(1-e^2)^3}(1+e\cos f)^2$$
$$\times [(2\beta + \gamma - 3\alpha) + (\gamma + 2)e^2 + 2(\beta - 2\alpha)e\cos f$$
$$- (2\gamma + 2 + \alpha)e^2 \cos^2 f] \tag{4.2.11a}$$

$$T = 2\frac{\mu}{c^2} \frac{\mu}{a^3(1-e^2)^3}(1+e\cos f)^3(\gamma + 1 - \alpha)\,e\sin f \tag{4.2.11b}$$

$$W = 0\ . \tag{4.2.11c}$$

Here $W = 0$ indicates that, due to the post-Newtonian conservation of angular momentum, the motion is restricted to a plane. This also implies that,

$$\Delta I = 0 \tag{4.2.12a}$$
$$\Delta\Omega = 0\ . \tag{4.2.12b}$$

Integration of (4.2.7) over the time variable t or the true anomaly f with

$$dt = \frac{1}{n}\left(\frac{r}{a}\right)^2 \frac{1}{\sqrt{1-e^2}}\,df$$

gives :†

$$\Delta a = \frac{\mu e}{c^2(1-e^2)^2}\left\{\left[(10\alpha - (6\gamma + 4\beta + 4)) + e^2\left(\frac{3}{2}\alpha - (4+2\gamma)\right)\right]\cos f\right.$$
$$\left.+ (4\alpha - (2\gamma + 2 + \beta))\,e\cos 2f + \frac{\alpha}{2}e^2 \cos 3f\right\}\Big|_{t_0}^{t} \tag{4.2.12c}$$

$$\Delta e = \frac{\mu}{c^2 a(1-e^2)}\left\{\left[(3\alpha - 2\beta - \gamma) + e^2\left(\frac{11}{4}\alpha - 4 - 3\gamma\right)\right]\cos f\right.$$
$$\left.+ \left(2\alpha - \gamma - 1 - \frac{1}{2}\beta\right)e\cos 2f + \frac{\alpha}{4}e^2 \cos 3f\right\}\Big|_{t_0}^{t} \tag{4.2.12d}$$

$$\Delta\omega = \frac{\mu}{c^2 a(1-e^2)}\left\{(2\gamma + 2 - \beta)f + \left[\frac{3\alpha - \gamma - 2\beta}{e} + \left(\gamma + \frac{1}{4}\alpha\right)e\right]\sin f\right.$$
$$\left.+ \left(2\alpha - \gamma - 1 - \frac{1}{2}\beta\right)\sin 2f + \frac{\alpha}{4}e\sin 3f\right\}\Big|_{t_0}^{t} \tag{4.2.12e}$$

$$\Delta\epsilon = (1 - \sqrt{1-e^2})\,\Delta\omega + \frac{\mu}{c^2 a\sqrt{1-e^2}}\left[(2\gamma + 4)\sqrt{1-e^2}E\right.$$
$$\left.+ 2(3\alpha - 2\gamma - 2\beta - 2)f + 2(\alpha + 2\gamma + 2)\,e\sin f\right]\Big|_{t_0}^{t} \tag{4.2.12f}$$

---

† These results can also be obtained by the substitutions $\sigma \to \beta + \gamma - \alpha$, $\beta \to (\gamma + \alpha)/2$, $\lambda \to \gamma + 1 - \alpha$ in the results of Brumberg (1972), Chapter 3.

$$\int_{t_0}^{t} \Delta n \, dt = \frac{3\mu}{c^2 a} \left\{ -(\gamma+2)E + \frac{(2\gamma+\beta+2-2\alpha)}{\sqrt{1-e^2}} f - \alpha \frac{e \sin f}{\sqrt{1-e^2}} \right.$$

$$+ \left[\alpha(1-e^2) \left(\frac{a}{r_0} \right)^3 + (\alpha - 2\gamma - \beta - 2) \left(\frac{a}{r_0} \right)^2 \right.$$

$$+ \left. \left. (\gamma+2)\frac{a}{r_0} \right] M \right\} \Bigg|_{t_0}^{t} \qquad (4.2.13)$$

with

$$\Delta M = \Delta\epsilon - \Delta\omega + \int_{t_0}^{t} \Delta n \, dt \ ,$$

where E denotes the *eccentric anomaly*.

Figure 4.2 shows the perturbations of orbital elements a, e, ω and M for LAGEOS (geocentric standpoint!) for one revolution in the spherical Einstein PN field ($\beta = \gamma = 1$) of the Earth for three ($\alpha = 0, 1, 2$) systems of coordinates. Here, $\alpha = 0$ denotes the standard PN coordinates (spatially isotropic) and $\alpha = 1$ the spatial standard coordinate system.

One sees that the mean anomaly M, the argument of the perigee ω and the numerical eccentricity e show only small short periodic perturbations for $\alpha = 1$. This is related to the fact that in this case the mean motion of the satellite is given by

$$n = n_0 \left(1 - \frac{3}{2}(1-\alpha)\frac{\mu}{c^2 a} \right) \qquad (4.2.14)$$

with $n_0^2 a^3 = \mu$. Accordingly, in particular, $n = n_0$ for $\alpha = 1$, since in that case the circumference of a circle with radius r is given by $2\pi r$. According to this it is obvious that in the post-Newtonian framework the orbital elements only play the role of coordinates and depend upon the chosen gauge (i.e. upon α).

In practice one usually adopts standard PN coordinates (spatially iso- tropic, $\alpha = 0$, i.e. *no* spatial standard coordinates with $\alpha = 1$) for PPN ephemerides. Of course, the dynamics e.g. of a satellite can equally well be described in different coordinates. In that case one has to bear in mind that the measuring process (laser ranging measurement etc.) has to be described in the *same* coordinate system in order to ensure that the observables are correctly described as coordinate independent quantities.

The spherical PN field of the Earth according to (4.2.12e) leads to an (Einstein) secular perihelion shift of:

$$\langle \Delta\omega \rangle = \frac{(2\gamma - \beta + 2)\mu}{c^2 a(1-e^2)} n_0 t \ . \qquad (4.2.15)$$

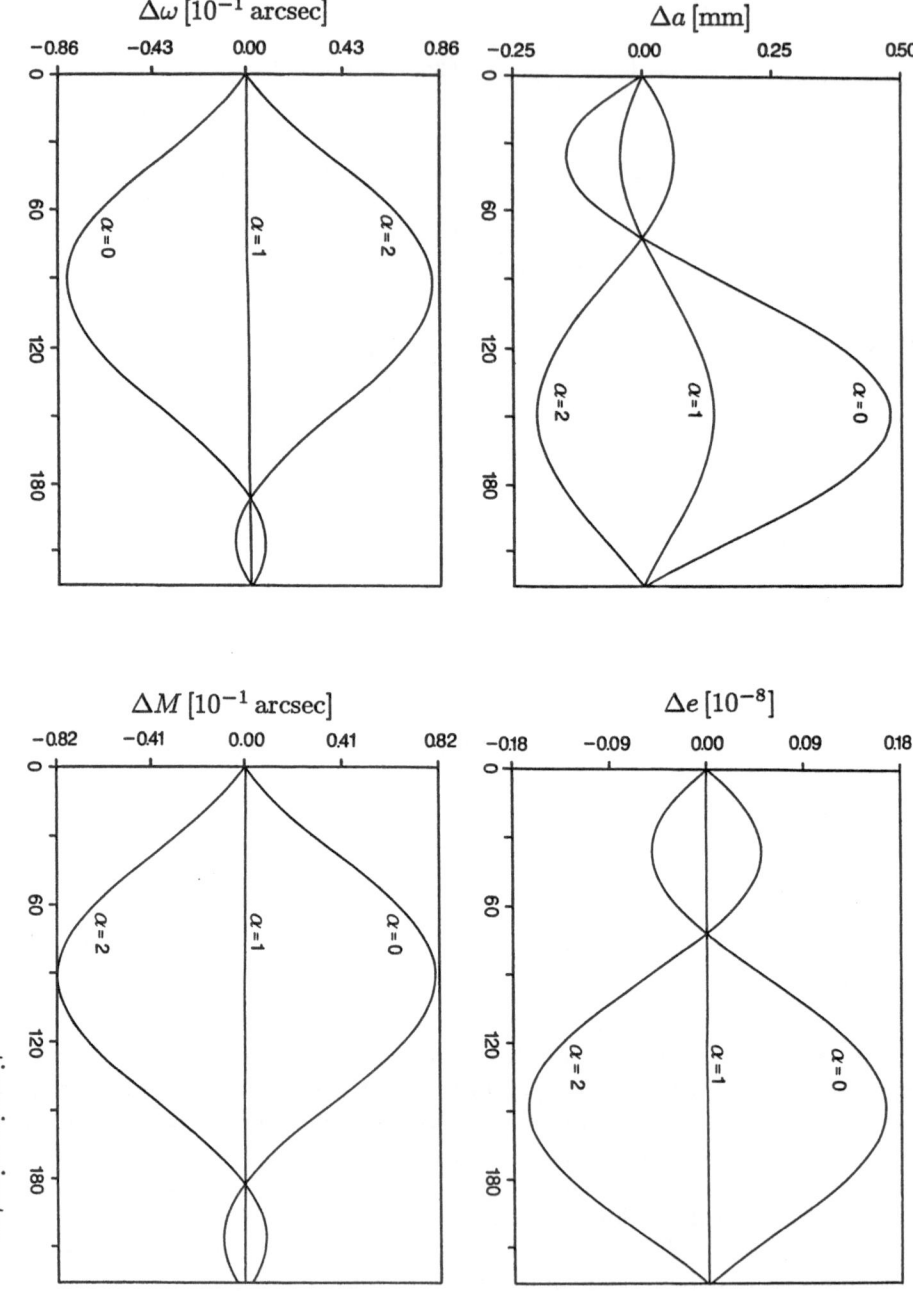

Fig. 4.2. Perturbations of orbital elements of LAGEOS for one revolution in the spherical Einstein PN field ($\beta = \gamma = 1$) of the Earth and for three different ($\alpha = 0, 1, 2$) coordinate systems

4.2.2 The Gravito-Magnetic Field of the Earth

According to the mass current and the gravito-magnetic field (g_{0i}) (cf. (4.1.3)) produced thereby a model Earth uniformly rotating about the z-axis leads to a perturbing (Lense-Thirring) acceleration of the form

$$\mathbf{a}_{\text{LT}} = \frac{(\gamma + 1)}{2} \mathbf{v} \times (\nabla \times \boldsymbol{\zeta}) \tag{4.2.16}$$

with

$$\boldsymbol{\zeta} = -4\mathbf{V} \equiv -2G \frac{\mathbf{J}_\oplus \times \mathbf{x}}{r^3} \,,$$

which is directly related to the Lense-Thirring precession of inertial axes w.r.t. fixed star oriented (Copernican) axes (cf. (3.3.73)), since the angular momentum vector of the satellite orbit can be considered as an inertial, Fermi-transported vector.

Also for the Lense-Thirring acceleration a perturbational theoretical treatment of the orbital elements is suggested. A decomposition of the perturbing acceleration (4.2.16) according to (4.2.5) yields because of

$$\mathbf{a}_{\text{LT}} = \frac{K}{r^3} \left(\begin{array}{c} [(x^2 + y^2 - 2z^2)/r^2]\, \dot{y} + 3(yz/r^2)\, \dot{z} \\ -[(x^2 + y^2 - 2z^2)/r^2]\, \dot{x} - 3(xz/r^2)\, \dot{z} \\ 3(z/r)[(x\dot{y} - y\dot{x})/r] \end{array} \right) \tag{4.2.17}$$

with

$$K = (\gamma + 1)\frac{GJ_\oplus}{c^2} = \frac{4}{5}\frac{(\gamma + 1)}{2}\frac{\mu \Omega_\oplus R_\oplus^2}{c^2}$$

and $(h^2 = \mu p; u = \omega + f)$:

$$S = Kh \frac{\cos I}{r^4} \tag{4.2.18a}$$

$$T = -Kh \frac{e \cos I \sin f}{pr^3} \tag{4.2.18b}$$

$$W = Kh \frac{\sin I}{r^4} \left(2\sin u + \frac{er \sin f \cos u}{p} \right) \,. \tag{4.2.18c}$$

For the changes of the orbital elements one then finds to first order (Lense et al. 1918) $(\xi = K/hp)$:

$$\Delta a = 0 \tag{4.2.19a}$$

$$\Delta e = -\xi \cos I (1 - e^2) \cos f \tag{4.2.19b}$$

$$\Delta I = -\xi \sin I/2 \left[\cos 2u + 2e \cos f \cos^2 u \right] \tag{4.2.19c}$$

$$\Delta \varpi = -\xi \cos I \left[2f + \frac{1 + e^2}{e} \sin f \right] + 2\sin^2 \frac{I}{2} \Delta \Omega \tag{4.2.19d}$$

$$\Delta\Omega = \xi \left[f - \frac{1}{2}\sin 2u + e(\sin f - \frac{1}{2}\sin 2u \cos f) \right] \qquad (4.2.19e)$$

$$\Delta\epsilon = -2(1-e^2)^{1/2}\xi\cos I(f + e\sin f) + \frac{e^2}{1+\sqrt{1-e^2}}\Delta\varpi$$

$$+ 2(1-e^2)^{1/2}\sin^2\frac{I}{2}\Delta\Omega \ . \qquad (4.2.19f)$$

For the secular perturbations of ϖ, Ω and ϵ one finds from (4.2.19d,e,f) that

$$\langle\Delta\Omega\rangle = \frac{\xi}{(1-e^2)^{3/2}}n_0 t \qquad (4.2.20a)$$

$$\langle\Delta\omega\rangle = -3\cos I\langle\Delta\Omega\rangle \qquad (4.2.20b)$$

and

$$\langle\Delta\varpi\rangle = \langle\Delta\epsilon\rangle = (1 - 3\cos I)\langle\Delta\Omega\rangle \ . \qquad (4.2.20c)$$

4.2.3 The Post-Newtonian Quadrupole

From the expression (4.2.1) for the post-Newtonian satellite acceleration in the gravitational field of the Earth one would expect that, in addition to the spherical PN field and the gravito-magnetic field of the Earth, essentially only the nonspherical part of the PN field of the Earth contributes to the relativistic perturbations of the satellite's orbital motion. We can absorb the potential Φ in U (e.g. Weinberg 1972), and obtain for the satellite acceleration without the Lense-Thirring part in the practically stationary field of the Earth:

$$\frac{d\mathbf{v}}{dt} = \nabla U + \frac{1}{c^2}[-(\beta+\gamma)\nabla U^2 - 2(\gamma+1)\mathbf{v}(\mathbf{v}\cdot\nabla)U + \gamma\mathbf{v}\nabla U] \ . \qquad (4.2.21)$$

If we consider only those PN perturbations that are induced by the quadrupole moment (J_2) of the Earth then the acceleration to first order in J_2 reduces to:

$$\mathbf{a}_Q \simeq \frac{1}{c^2}[-2(\beta+\gamma)\nabla(U_0 U_2) - 2(\gamma+1)\mathbf{v}(\mathbf{v}\cdot\nabla)U_2 + \gamma\mathbf{v}^2\nabla U_2]$$

$$\equiv \mathbf{a}_1 + \mathbf{a}_2 + \mathbf{a}_3 \qquad (4.2.22)$$

with

$$U_0 = \frac{\mu}{r}$$

$$U_2 = -\frac{1}{2}\frac{\mu}{r}J_2\left(\frac{R}{r}\right)^2\left(3\frac{z^2}{r^2} - 1\right) \ .$$

One finds:

$$\mathbf{a}_1 = 2(\beta+\gamma)\left(\frac{\mu}{c^2 r}\right)J_2\left(\frac{R}{r}\right)^2\begin{pmatrix} (x/r)(2 - (9z^2/r^2)) \\ (y/r)(2 - (9z^2/r^2)) \\ (z/r)(5 - (9z^2/r^2)) \end{pmatrix}\frac{\mu}{r^2} \qquad (4.2.23)$$

$$\mathbf{a}_2 = 3(\gamma + 1) \left(\frac{\mu}{c^2 r}\right) J_2 \left(\frac{R}{r}\right)^2 \left[\left(1 - 5\frac{z^2}{r^2}\right) \frac{(xv_x + yv_y)}{r} + \left(3 - 5\frac{z^2}{r^2}\right) \frac{zv_z}{r}\right] \frac{\mathbf{v}}{r}$$

$$(4.2.24)$$

$$\mathbf{a}_3 = -\frac{3}{2}\gamma \left(\frac{\mu}{c^2 r}\right) J_2 \left(\frac{R}{r}\right)^2 \begin{pmatrix} (x/r)(1 - (5z^2/r^2)) \\ (y/r)(1 - (5z^2/r^2)) \\ (z/r)(3 - (5z^2/r^2)) \end{pmatrix} \frac{\mathbf{v}^2}{r^2} .$$

$$(4.2.25)$$

The decomposition of perturbing acceleration $\mathbf{a}_Q$, as well as the resulting expressions for the perturbations of orbital elements can be found in Soffel et al. (1988b). It should be noted that the second order mixed perturbations due to the Newtonian quadrupole field and the Schwarzschild field are of the same order ($\propto J_2 c^{-2}$) as the PN-perturbations arising from $\mathbf{a}_Q$ (4.2.22). These mixed perturbations have not yet been treated in the literature.

4.3 Lunar Motion

The Moon moves in an approximately elliptical orbit inclined at about five degrees to the plane of the ecliptic. The mean values of the semi-major axis a the eccentricity e and the inclination I are: $a = 384\,400$ km, $e = 0.054\,90$ and $I = 5°09'$. Because of solar perturbations all three elements are subject to periodic variations about these values. In particular, the eccentricity varies from 0.044 to 0.067 while the inclination oscillates between $4°58'$ and $5°19'$ (Roy 1978). Usually, the position of the Moon is described by longitude λ, latitude β w.r.t. the ecliptic, and the distance or parallax. Via the true obliquity of the ecliptic ϵ longitude and latitude of the Moon are related to right ascension and declination by

$$\cos \delta \cos \alpha = \cos \beta \cos \lambda$$
$$\cos \delta \sin \alpha = \cos \beta \sin \lambda \cos \epsilon - \sin \beta \sin \epsilon$$
$$\sin \delta = \cos \beta \sin \lambda \sin \epsilon + \sin \beta \cos \epsilon .$$

An analytical theory describing these quantities which are directly related to observational data with great precision is the one developed by Brown (e.g. Brown 1960, Plummer 1960). In Brown's theory longitude, latitude and parallax (distance) are expressed as Poisson series where the arguments of the trigonometrical functions are linear combinations of *fundamental angles*; some of them have already been introduced in Section 3.2. As far as the problem of Sun, Moon and Earth is concerned these fundamental arguments are:

$$L = \text{mean longitude of the Moon}$$
$$L' = \text{mean longitude of the Sun}$$
$$\varpi = \text{mean longitude of the Moon's perigee}$$
$$\varpi' = \text{mean longitude of the Sun's perigee}$$
$$\Omega = \text{mean longitude of the Moon's node}$$

and they occur only as the differences

$$l = L - \varpi$$
$$l' = L' - \varpi'$$
$$F = L - \Omega$$
$$D = L - L' \ .$$

Precise values for these fundamental angles referred to the epoch JD 2451545.0 = 2000 January 1.5 are given in Table 4.1 (Merit Standards 1983). Such high precision usually is needed for the theory of nutations.

Table 4.1 Fundamental Angles for the Problem of Sun, Moon and Earth

$$l = 134° \ 57' \ 46''.733 + (1325^r + 198° \ 52' \ 02''.633) \ T$$
$$+ \ 31''.310 \ T^2 + 0''.064 \ T^3$$

$$l' = 357° \ 31' \ 39''.804 + (0099^r + 359° \ 03' \ 01''.224) \ T$$
$$- \ 00''.577 \ T^2 - 0''.012 \ T^3$$

$$F = 093° \ 16' \ 18''.877 + (1342^r + 082° \ 01' \ 03''.137) \ T$$
$$- \ 13''.257 \ T^2 + 0''.011 \ T^3$$

$$D = 297° \ 51' \ 01''.307 + (1236^r + 307° \ 06' \ 41''.328) \ T$$
$$- \ 06''.891 \ T^2 + 0''.019 \ T^3$$

$$\Omega = 125° \ 02' \ 40''.280 + (0005^r + 134° \ 08' \ 10''.539) \ T$$
$$+ \ 07''.455 \ T^2 + 0''.008 \ T^3$$

T is measured in Julian centuries of 36 525 days of 86 400 seconds of Dynamical Time since J2000.0 and the superscript r indicates complete revolutions of 360°

The quantity easily accessible to observation is the lunar longitude λ. The first few terms in the expression for the Moon's longitude are:

$$\lambda = n(t - t_0) + 377' \ \sin l + 13' \ \sin 2l + 76' \ \sin(2D - l)$$
$$+ 40' \ \sin 2D - 11' \ \sin l' + \cdots \ . \tag{4.3.1}$$

In this expression the terms l and $2l$ are ordinary elliptic two-body terms proportional to the eccentricity of the lunar orbit. The $(2D - l)$-term is called *evection*; it is due to the variation in the eccentricity of the orbit caused by the solar tidal forces. The evection has a period of ~ 31.8 days and was already known to Ptolemy. The $2D$-term is the *variation*, caused by the variation of the solar force in course of a synodic month of 29.53059 days. It has a period of half a synodic month and was first discovered by Tycho Brahe. Finally, the l'-term with a period of one (anomalistic) year is called the *annual equation*; it is caused by the annual variation of the solar tidal forces due to the eccentricity of the Earth-Moon orbit about the Sun. To lowest order the Hill-Brown theory gives the longitude of the Moon as:†

$$\lambda = n(t - t_0) + e \sin l + \frac{5}{16} e^2 \sin 2l + \frac{15}{8} me \sin(2D - l)$$
$$+ \frac{11}{8} m^2 \sin 2D - 3e'm \sin l' + \cdots . \qquad (4.3.2)$$

Here, 'e' is about twice the numerical eccentricity of the Moon's orbit $e \simeq 0.11$, $m = n'/(n - n') \simeq 0.08$ and $e' \simeq 0.017$, where the prime refers to the solar orbit. The principal term in the longitude depending upon a' (the "solar parallax") reads

$$-\frac{15}{8} \alpha m \sin D ,$$

where $\alpha \simeq 0.00257$ is roughly given by a/a'. The full D-term has an amplitude of about $-2'$.

As early as 1916, soon after Einstein's paper on the foundations of general relativity (Einstein 1915) appeared in the literature, de Sitter (1916) had worked out the expected general relativistic contributions to the secular motion of the lunar perigee and node (see also Chazy 1928, 1930, Eddington 1975). Later Fokker (1921, 1965) showed that this (geodetic) precession of the lunar orbit applies to all gyroscopes moving with the Earth-Moon system about the Sun and that the origin of this precession is a small difference between proper orbital frequency and orbital frequency as determined by an asymptotic observer at rest w.r.t. the barycenter. In 1958 Brumberg (1958) presented a solution of the restricted post-Newtonian 3-body problem based on the Hill-Brown method with circular motion for the Earth-Moon system about the Sun. Baierlein (1967) essentially verified Brumberg's calculation and took the eccentricity of the Earth's orbit into account. These calculations were extended by Krogh et al. (1968) to cover the predictions from Brans-Dicke theory. More recently, post-Newtonian theories of great accuracy have been formulated by Brumberg et al. (1982) and by Lestrade et al. (1982). The motion of the Moon in a Fermi-frame moving with the Earth-Moon barycenter has been studied by Mashhoon (1985) and Soffel et al. (1986).

---

† A first impression of the Hill-Brown theory can be obtained from A.3 of the Appendix, where the variational (or D) terms are considered. For more information the reader is referred to Brown (1960) and Plummer (1960).

Brumberg's (1958) post-Newtonian Hill-Brown theory for the lunar motion in the Einstein case ($\beta = \gamma = 1$) is concerned with the motion of the Moon in barycentric PN coordinates; i.e. his expressions for the lunar distance ("radius vector") r, longitude λ and $b = z/r$ value (or latitude β; $\tan \beta = b$) as function of coordinate time are *coordinate quantities* in a barycentric frame of reference. Brumberg's expressions read (Brumberg 1958, 1972):

$$
\begin{aligned}
\frac{r}{a_0} = 1 + \sigma &\left[-\frac{9}{4} + m - \frac{47}{32}m^2 - \frac{3}{16}m^3 \right.\\
&\left. - \frac{\alpha^2}{m^2}(1 + 2m) \right] + \left(\frac{5}{16}\alpha m + \cdots - \sigma\frac{7}{2}\alpha \right) \cos D \\
&+ \left[-m^2 - \frac{7}{6}m^3 + \cdots + \sigma \left(\frac{1}{4} + \frac{17}{4}m^2 + \frac{137}{24}m^3 \right) \right] \cos 2D \\
&+ \left(\cdots - \sigma\frac{173}{128}\alpha m \right) \cos 3D + \left[\cdots + \sigma \left(\frac{7}{32}m^2 \right. \right.\\
&\left. + \frac{19}{48}m^3 \right) \bigg] \cos 4D + \cdots + e \left\{ \left[-\frac{1}{2} + \frac{3}{8}m^2 \right. \right.\\
&+ \frac{1311}{2^9}m^3 + \cdots + \sigma \left(\frac{9}{8} - \frac{233}{2^8}m - \frac{1627}{2^{10}}m^2 - \frac{264877}{2^{14}}m^3 \right.\\
&\left. + \frac{1}{2}\frac{\alpha^2}{m^2}(1 + 2m) \right) \bigg] \cos l + \left[-\frac{17}{2^5}m^2 - \frac{49}{3 \cdot 2^5}m^3 \right.\\
&+ \cdots + \sigma \left(\frac{3}{2^4} + \frac{295}{2^7}m^2 + \frac{6961}{3 \cdot 2^{10}}m^3 \right) \bigg] \cos(2D + l) \\
&+ \left[\cdots + \sigma \left(\frac{85}{2^8}m^2 + \frac{419}{3 \cdot 2^8}m^3 \right) \right] \cos(4D + l) \\
&+ \left[-\frac{15}{2^4}m - \frac{95}{2^6}m^2 - \frac{12169}{3 \cdot 2^{10}}m^3 + \cdots + \sigma \left(-\frac{5}{2^4} \right.\right.\\
&\left. + \frac{165}{2^6}m + \frac{7451}{2^{11}}m^2 \right) \bigg] \cos(2D - l) + \left[-\frac{255}{2^8}m^3 \right.\\
&+ \cdots + \sigma \left(\frac{45}{2^7}m + \frac{561}{2^9}m^2 + \frac{64817}{2^{13}}m^3 \right) \bigg] \cos(4D - l) \\
&+ \left(\cdots + \sigma\frac{1275}{2^{11}}m^3 \right) \cos(6D - l) + \cdots \bigg\} + \cdots
\end{aligned}
\tag{4.3.3}
$$

for the coordinate distance,

$$
\begin{aligned}
\lambda = n(t - t_0) &+ \left[-\frac{15}{8}\alpha m + \cdots + \sigma \left(1 + \frac{13}{2}m \right) \frac{\alpha}{m} \right] \sin D \\
&+ \left[\frac{11}{8}m^2 + \frac{13}{6}m^3 + \cdots + \sigma \left(-\frac{1}{4} - \frac{11}{4}m^2 \right) \right.
\end{aligned}
$$

$$- \frac{91}{12}m^3 \Big) \Big] \sin 2D + \Big(\cdots + \sigma \frac{65}{32}\alpha m \Big) \sin 3D + \Big[\cdots$$

$$+ \sigma \Big(-\frac{11}{32}m^2 - \frac{13}{24}m^3 \Big) \Big] \sin 4D + \cdots + e \Big\{ \Big[1 - \frac{75}{2^8}m^3$$

$$+ \cdots + \sigma \Big(\frac{15}{2^7}m + \frac{325}{2^9}m^2 + \frac{60953}{3 \cdot 2^{13}}m^3 \Big) \Big] \sin l$$

$$+ \Big[\frac{17}{2^4}m^2 + \frac{67}{3 \cdot 2^4}m^3 + \cdots + \sigma \Big(-\frac{1}{4} - \frac{17}{8}m^2 \Big.$$

$$- \frac{27901}{3 \cdot 2^{11}}m^3 \Big) \Big] \sin(2D + l) + \Big[\cdots + \sigma \Big(-\frac{39}{2^6}m^2 \Big.$$

$$- \frac{57}{2^6}m^3 \Big) \Big] \sin(4D + l) + \Big[\frac{15}{2^3}m + \frac{203}{2^5}m^2 + \frac{25849}{3 \cdot 2^9}m^3$$

$$+ \cdots + \sigma \Big(\frac{1}{4} - \frac{15}{2^4}m - \frac{9347}{2^{10}}m^2 \Big) \Big] \sin(2D - l)$$

$$+ \Big[\frac{255}{2^7}m^3 + \cdots + \sigma \Big(-\frac{15}{2^5}m - \frac{159}{2^7}m^2 - \frac{17707}{2^{11}}m^3 \Big) \Big] \sin(4D - l)$$

$$+ \Big(\cdots - \sigma \frac{585}{2^9}m^3 \Big) \sin(6D - l) + \cdots \Big\} + \cdots \tag{4.3.4}$$

for the longitude and

$$b = \frac{z}{r} = 2K \Big\{ \Big[1 - \frac{3}{2^4}m^3 + \cdots + \sigma \Big(\frac{3}{2^6}m + \frac{3}{2^8}m^2 \Big.$$

$$+ \frac{2255}{2^{12}}m^3 \Big) \Big] \sin F + \Big[\frac{11}{2^4}m^2 + \frac{13}{12}m^3 + \cdots$$

$$+ \sigma \Big(-\frac{1}{8} - \frac{11}{8}m^2 - \frac{8857}{3 \cdot 2^9}m^3 \Big) \Big] \sin(2D + F)$$

$$+ \Big[\cdots + \sigma \Big(-\frac{33}{2^7}m^2 - \frac{13}{2^5}m^3 \Big) \Big] \sin(4D + F)$$

$$+ \Big[\frac{3}{8}m + \frac{13}{2^5}m^2 + \frac{1133}{3 \cdot 2^9}m^3 + \cdots + \sigma \Big(\frac{1}{8} - \frac{3}{2^4}m \Big.$$

$$- \frac{53}{2^5}m^2 \Big) \Big] \sin(2D - F) + \Big[\frac{33}{2^7}m^3 + \cdots + \sigma \Big(-\frac{3}{2^6}m \Big.$$

$$+ \frac{31}{2^8}m^2 - \frac{5725}{3 \cdot 2^{12}}m^3 \Big) \Big] \sin(4D - F)$$

$$+ \Big(\cdots - \sigma \frac{99}{2^{10}}m^3 \Big) \sin(6D - F) + \cdots \Big\} + \cdots . \tag{4.3.5}$$

Here $\bar{a}_0$ is given by

$$\bar{a}_0 = a_0 \Big(1 - \frac{1}{6}m^2 + \frac{1}{3}m^3 + \cdots \Big)$$

with $n_0^2 a_0^3 = GM_\oplus$. The mean motion of the Earth-Moon system (or the Sun) is given by

$$n' = \left(\frac{GM_\odot}{a'^3}\right)^{1/2}\left(1 - \frac{3}{2}\sigma\right) \tag{4.3.6}$$

and

$$\sigma \equiv \frac{Gm_\odot}{c^2 a'} \simeq 10^{-8} \tag{4.3.7}$$

characterizes the relativistic terms. The constant $K \simeq 0.045$ in the expression for z/r differs little from $0.5 \tan I$.

For the secular motion of the perigee and node one finds (Brumberg 1958, 1972):

$$\frac{1}{n}\langle\dot{\varpi}\rangle = \frac{3}{4}m^2 + \frac{177}{32}m^3 + \cdots + \sigma\left[\frac{3}{2}m - \frac{3}{2}m^2 - \frac{957}{64}m^3 + 3\frac{\alpha^2}{m^2}(1 + 2m)\right]$$

$$\tag{4.3.8}$$

$$\frac{1}{n}\langle\dot{\Omega}\rangle = -\frac{3}{4}m^2 + \frac{57}{32}m^3 + \cdots + \sigma\left[\frac{3}{2}m - \frac{3}{2}m^2 - \frac{21}{64}m^3\right] \ . \tag{4.3.9}$$

For the simplified problem of vanishing eccentricities and solar parallax a PPN-Hill-Brown theory for the variational terms is formulated in A.3 of the Appendix. In standard barycentric PN coordinates one finds for the coordinate distance and longitude:

$$\frac{r}{a_0} = 1 - m^2 \cos 2D + \cdots$$

$$+ \sigma\left\{-\frac{2}{3}\gamma - \frac{4}{3}\beta - \frac{1}{4} + \frac{1}{3}(2\gamma + 1)m - \left(\frac{10}{3}\gamma - 3\beta + \frac{109}{96}\right)m^2 + \cdots\right.$$

$$\left. + \left[\frac{1}{4} + \left(\frac{1}{4} + 2\gamma + 2\beta\right)m^2\right]\cos 2D + \frac{7}{32}m^2 \cos 4D + \cdots\right\} \tag{4.3.10a}$$

and

$$\lambda = n(t - t_0) + \frac{11}{8}m^2 \sin 2D + \cdots$$

$$+ \sigma\left\{\left[-\frac{1}{4} - \frac{11}{12}(2\gamma + \beta)m^2\right]\sin 2D - \frac{11}{32}m^2 \sin 4D + \cdots\right\} \tag{4.3.10b}$$

and the mean motion of the Sun is given by:

$$n' = \left(\frac{GM_\odot}{a'^3}\right)^{1/2}\left[1 - \frac{1}{2}(2\beta + \gamma)\sigma\right] \ . \tag{4.3.11}$$

Table 4.2 Relativistic Perturbations of the Earth-Moon Coordinate Distance r/a₀.

D	l	l'	$N \times 10^1$	$R \times 10^8$	$\beta \times 10^8$	$\gamma \times 10^8$	$E \times 10^8$
0	0	0	9.9912	−0.2273	−1.3132	−0.6086	−2.1491
2	0	0	−0.0717	0.2486	0.0141	0.0145	0.2772
4	0	0	−0.0002	0.0016	0.0001	0.0001	0.0018
0	1	0	−0.5438	0.0060	0.0707	0.0327	0.1094
2	−1	0	−0.0968	−0.0378	0.0206	0.0093	−0.0079
2	1	0	−0.0041	0.0204	0.0008	0.0008	0.0221
4	−1	0	−0.0008	0.0040	0.0002	0.0002	0.0045
0	0	1	0.0012	−0.0583	0.0665	−0.0333	−0.0251
2	0	−1	−0.0051	0.0167	−0.0133	0.0056	0.0089
2	0	1	0.0007	−0.0062	0.0114	−0.0030	0.0021
1	0	0	0.0029	0.0841	−0.1239	0.0246	−0.0152

The various columns indicate various contributions to the Poisson amplitudes. N: Newtonian term, R: PPN parameter-free part, β, γ: amplitude proportional to β, γ, E: full relativistic amplitude in the Einstein case ($E = R + \beta + \gamma$).

Table 4.3 Relativistic Perturbations of the Moon's Longitude in arcsec.

D	l	l'	$N \times 10^{-4}$	$R \times 10^3$	$\beta \times 10^3$	$\gamma \times 10^3$	$E \times 10^3$
2	0	0	0.2106	−0.5104	−0.0138	−0.0305	−0.5547
4	0	0	0.0008	−0.0052	−0.0001	−0.0003	−0.0056
0	1	0	2.2641	0.0121	0.0001	0.0001	0.0123
2	−1	0	0.4606	0.0814	−0.0368	−0.0229	0.0217
2	1	0	0.0175	−0.0559	−0.0011	−0.0025	−0.0595
4	−1	0	0.0035	−0.0106	−0.0005	−0.0007	−0.0118
0	0	1	−0.0655	1.8731	−3.6707	1.0139	−0.7838
2	0	−1	0.0152	−0.0402	0.0406	−0.0153	−0.0150
2	0	1	−0.0022	0.0182	−0.0345	0.0091	−0.0071
1	0	0	−0.0129	−0.2943	0.5379	−0.1155	0.1281

Table 4.4 Relativistic Perturbations of the Moon's Latitude in arcsec.

D	F	$N \times 10^{-4}$	$R \times 10^5$	$\beta \times 10^5$	$\gamma \times 10^5$	$E \times 10^5$
0	1	1.8516	4.6799	0.0036	0.0059	4.6894
2	−1	0.0618	2.6523	−0.3835	−0.3391	1.9297
2	1	0.0094	−2.2669	−0.0616	−0.1367	−2.4652

Numerical values of the various post-Newtonian perturbations of the coordinate distance, longitude and latitude of the Moon's orbit are given in Tabs. 4.2–4.4 (data taken from Brumberg et al. 1982).

As we can see, the dominant relativistic oscillation of the coordinate distance has an amplitude of $\sim 100\,\mathrm{cm}$ and a period of half a synodic month.

The dominant contribution to the secular advances of the perigee and node (4.3.8, 4.3.9) results from the *geodetic precession*. As outlined in A3 the Einstein-Infeld-Hoffmann equations of motion for the lunar orbit due to relativistic kinematics yield a perturbing acceleration in the Earth-Moon system that is roughly constant and of the form

$$R_{GP} = \Omega_{GP}\, na^2 \sqrt{1 - e^2}\, \cos I \ , \tag{4.3.12}$$

where

$$\Omega_{GP} = -\frac{(2\gamma + 1)}{2}\sigma n' \tag{4.3.13}$$

represents the angular velocity of geodetic precession. This results in a secular advance of the longitude of perigee, the node and the mean longitude at the epoch with

$$\langle \dot{\varpi} \rangle = \langle \dot{\Omega} \rangle = -\Omega_{GP} \tag{4.3.14a}$$

and

$$\langle \dot{\varepsilon} \rangle = -\left(1 + 3\sqrt{1 - e^2}\, \cos I\right)\Omega_{GP} \ . \tag{4.3.14b}$$

Table 4.5 taken from Brown's lunar theory (Roy 1978) displays the various Newtonian contributions to the secular motion of the Moon's longitude of perigee ϖ and node Ω.

Table 4.5 Motion of the Lunar Perigee and Node in arcsec/y

Mean motion of	Perigee	Node
Principal solar action	+146 426.92	−696 72.04
Mass of the Earth	−0.68	+ 0.19
Direct planetary action	+2.69	−1.42
Indirect planetary action	−0.16	+ 0.05
Figure of the Earth	+ 6.41	−6.00
Figure of the Moon	+ 0.03	−0.14
Sum	+146 435.21	−69 679.36

The general relativistic contributions to $\langle \dot{\varpi} \rangle$ and $\langle \dot{\Omega} \rangle$ amount to ($\beta = \gamma = 1$; Brumberg et al. 1982):

$$\langle \dot{\varpi} \rangle = 1.728''/\text{century} \quad ; \quad \langle \dot{\Omega} \rangle = 1.901''/\text{century} \ .$$

Though the barycentric coordinate motion of the Moon is most useful for practical purposes it might be of interest to study the lunar motion in a Fermi frame comoving with the Earth-Moon barycenter. The EIH equations leading to this result read in instantaneous (not co-moving) geocentric coordinates (A3.16) ($M = M_\oplus$):

$$\ddot{x} - 2n'\dot{y} + GM\frac{x}{r^3} - 3n'^2 x = \sigma\left[-2\gamma n'\dot{y} - 2\delta_3 n'^2 x + 2\alpha_3 GM\frac{x}{r^3} + \frac{3}{2}GM\frac{xy^2}{r^5}\right]$$

$$\ddot{y} + 2n'\dot{x} + GM\frac{y}{r^3} = \sigma\left[+2\alpha_2 n'\dot{x} + 2\alpha_3 GM\frac{y}{r^3} + \frac{3}{2}GM\frac{y^3}{r^5}\right]$$

$$\ddot{z} + GM\frac{z}{r^3} + n'^2 x = \sigma\left[+ 2\alpha_3 GM\frac{z}{r^3} + \frac{3}{2}GM\frac{zy^2}{r^5}\right]$$

$$(4.3.15)$$

with $\alpha_2 = \gamma + 1$, $\delta_3 = 2\gamma + \beta$ and $\alpha_3 = 2\beta + \gamma$. We would now like to switch to proper coordinates (τ, X, Y, Z) of an observer co-moving with the barycenter of the Earth-Moon system. According to Section 3.3.3 this transition is given by ($\sigma_{+\epsilon} \equiv 1 + \epsilon\sigma$):

$$\tau = \sigma_{-3/2}t \quad ; \quad X = \sigma_\gamma x \quad ; \quad Y = \sigma_\gamma \sigma_{1/2}y \quad ; \quad Z = \sigma_\gamma z \ .$$

Physically this transformation represents nothing but a Lorentz transformation in y-direction (the $\sigma_{1/2}$-term) and a rescaling of lengths (the σ_γ-terms) due to the spatial metric. If the dot now represents the derivative w.r.t. proper time τ in the proper reference frame the EIH equation takes the form ($R^2 = X^2 + Y^2 + Z^2$):

$$\ddot{X} - 2n'\dot{Y} + \frac{G_L MX}{R^3} - 3n'^2 X = \sigma\left[(9 - 4\gamma - 2\beta)n'^2 X - 2(\gamma - 1)n'\dot{Y}\right]$$

$$\ddot{Y} + 2n'\dot{X} + \frac{G_L MY}{R^3} = \sigma\left[+ 2(\gamma - 1)n'\dot{X}\right]$$

$$\ddot{Z} + \frac{G_L MZ}{R^3} + n'^2 X = \sigma\left[- 3n'^2 Z \right]$$

$$(4.3.16)$$

where the "locally measured gravitational constant" G_L is given by

$$G_L = G\left(1 - \eta_N U_\odot/c^2\right) \tag{4.3.17}$$

with

$$\eta_N \equiv (4\beta - \gamma - 3) \ . \tag{4.3.18}$$

This location dependent value for G_L can be related with the Nordtvedt-effect (Section 4.7) or a break-down of the *strong equivalence principle* (Will 1981).

Equation (4.3.16) now represents nothing but the Jacobi equation that without gravitational effects from the Earth and $\eta_N = 0$ reads (cf. (3.3.55)):

$$\ddot{\mathbf{X}} - 2\mathbf{N} \times \dot{\mathbf{X}} + \mathbf{N} \times (\mathbf{N} \times \mathbf{X}) + \hat{\mathbf{K}} \cdot \mathbf{X} = 0 \tag{4.3.19}$$

with

$$K_{(i)(j)} = R_{(0)(i)(0)(j)} = R_{\mu\nu\rho\sigma}\, e^{\mu}_{(0)} e^{\nu}_{(i)} e^{\rho}_{(0)} e^{\sigma}_{(j)} \ .$$

The tetrad vectors $e_{(\alpha)}$ with induced proper coordinates $(\tau, \mathbf{X})$ are given by (3.3.84) with $\phi_F = 0$ ($e_{(r)} = e_{(X)}$ always points in radial direction) and the angular velocity $\mathbf{N}$ of $e_{(\alpha)}$ w.r.t. the Fermi tetrad is given by

$$\mathbf{N} = (0, 0, \sigma_{-(\gamma-1)}\, n') \ . \tag{4.3.20}$$

If the gravitational field of the Earth is taken into account (Section 3.4) the Jacobi equation takes the form:

$$\ddot{X} - 2N\dot{Y} + \frac{GMX}{R^3} - N^2 X = F_X$$
$$\ddot{Y} + 2N\dot{X} + \frac{GMX}{R^3} - N^2 Y = F_Y \tag{4.3.21}$$
$$\ddot{Z} + \frac{GMZ}{R^3} = F_Z$$

with

$$F_i = -K_{(i)(j)} X^j \ .$$

If the tidal matrix $K_{(i)(j)}$ is computed by means of (A1.10) and ((3.3.84); $\phi_F = 0$), one obtains:

$$K_{(i)(j)} = \mathrm{diag}\Big[-2n_0'^2 \Big(1 - \frac{4\gamma + 6\beta - 7}{2}\sigma\Big);$$
$$n_0'^2 \Big(1 - (3\gamma + 2\beta - 2)\sigma\Big); n_0'^2 \Big(1 - (\gamma + 2\beta - 3)\sigma\Big)\Big] \ , \tag{4.3.22}$$

which brings the Jacobi equation precisely into the form (4.3.16) for $\eta_N = 0$.

In the Einstein post-Newtonian case a Hill-Brown treatment of (4.3.16) reduces to solving the Hill-Brown equations (A3.29, A3.30) with the generalized potential

$$\Omega = \frac{\kappa}{R} + \frac{3}{8}m^2 (u + s)^2 (1 + \sigma) \tag{4.3.23}$$

with

$$\kappa \equiv \frac{GM}{(n - n')^2}$$

and

$$X \equiv \frac{(u + s)}{2} \quad ; \quad Y \equiv \frac{(u - s)}{2i} \ .$$

Let $D_\tau \equiv (n - n')(\tau - \tau_0)$ be the mean longitude of (Moon − Sun) w.r.t. Fermi axes (not w.r.t. fixed-star oriented axes). The oscillations in radius vector and longitude are then essentially given by $(1 + \sigma) \times$ (the Newtonian expression):

$$\frac{R}{\bar{a}_0} = 1 + \sigma \left[-\frac{1}{2}m^2 + \frac{13}{6}m^3 + \cdots \right]$$

$$+ \left[\left(-m^2 - \frac{7}{6}m^3 + \cdots \right) (1 + \sigma) \right] \cos 2D_\tau + \cdots \qquad (4.3.24)$$

$$\Lambda = n(\tau - \tau_0) + \left[\left(\frac{11}{8}m^2 + \frac{13}{6}m^3 + \cdots \right) (1 + \sigma) \right] \sin 2D_\tau + \cdots . \qquad (4.3.25)$$

This means that in the comoving proper frame the amplitude of the dominant relativistic range oscillation amounts to ~ 2 cm (Mashhoon 1985, Soffel et al. 1986). As discussed in A.3 in more detail the oscillation amplitude of about 1 m, mentioned above, represents nothing but the Lorentz contraction of the lunar orbit with full amplitude $\delta r / r \simeq \frac{1}{2}(v_\oplus/c)^2$, if the Moon is in 1st and 3rd quarter and vanishing amplitude for new and full Moon. Now, $(v_\oplus/c)^2 \simeq \sigma$ and the Lorentz contraction of the lunar orbit leads to oscillations in distance of the form $\delta r / r \simeq +\frac{1}{4}\sigma \cos 2D$.

We finally would stress again that the use of barycentric ephemerides (here for the lunar orbit) requires a careful treatment of the observables, such as observed laser travel time intervals. Both proper time intervals and proper spatial directions have to be related with our coordinates in a barycentric frame of reference (cf. Chapter 3).

4.4 Motion of the Planetary System

The relativistic "corrections" in the orbital motions of the planets and the barycenter of the Earth-Moon system have been extensively discussed by Lestrade (1981) and Lestrade et al. (1982). If such a discussion is restricted to the effects caused by the spherical post-Newtonian field of the Sun (as is here the case) we can take the results from A.2 to study the variations of orbital elements in first order perturbation theory. Lestrade et al. also compute effects of second order (leading to perturbing terms proportional to T^2, $T \sin(n\lambda_i - m\lambda_j)$, $T \cos(n\lambda_i - m\lambda_j)$) and thereby use the elements: a (semi-major axis), λ (mean longitude), defined by $d\lambda/dt = n_0 + d\epsilon/dt$ (ϵ is the mean longitude of the epoch, $n_0^2 a^3 = G(M_\odot + m)$), $k = e \cos(\omega + \Omega)$, $h = e \sin(\omega + \Omega)$, $p = \frac{1}{2}\sin I \sin \Omega$ und $q = \frac{1}{2}\sin I \cos \Omega$. These elements are written in form of Poisson series w.r.t. the mean longitudes, i.e.

$$\sigma = \sigma_0 + \Delta\sigma$$

$$= \sigma_0 + \sum_{n=0}^{\infty} (A_n \sin n\lambda + B_n \cos n\lambda) + b\,t ,$$

if σ denotes an arbitrary element and one restricts consideration to the first order. Lestrade et al. have computed the variations of orbital elements with the requirement that the accuracy of 5×10^{-12} (A.U. or rad) is maintained over an interval of 1000 years. In Lestrade et al. (1982) the expressions are given with a precision of 10^{-10} over a span of 100 years.

According to the post-Newtonian conservation of angular momentum (cf. A.2) the inclination and the ascending node are conserved quantities in the central, spherical PN-field, i.e. $p = p_0$, $q = q_0$. If the last expression for the other orbital elements is inserted into the perturbation equations one finds (Lestrade 1981):†

$$\Delta a(t) = \delta a + \frac{2\mu}{c^2(1-e_0^2)^2} \left\{ \left[(-2 - 3\gamma - 2\beta) + \left(\frac{9}{4} + \frac{13}{4}\beta + \frac{35}{8}\gamma \right) e_0^2 \right] \right. $$
$$\times [h_0 \, \sin\lambda + k_0 \, \cos\lambda]$$
$$\left. + \left(-3 - 4\gamma - \frac{5}{2}\beta \right) [2h_0 k_0 \, \sin 2\lambda + (k_0^2 - h_0^2) \, \cos 2\lambda] + \cdots \right\}$$

$$(4.4.1a)$$

$$\Delta\lambda(t) = \frac{\mu}{a_0(1-e_0^2)^2} \left\{ \left[\left(\frac{1 - \sqrt{1-e_0^2}}{e_0^2} \right) (-\gamma - 2\beta) + \left(16 + \frac{159}{8}\gamma + \frac{35}{4}\beta \right) \right. \right.$$
$$\left. + \left(-10 - \frac{103}{8}\gamma - \frac{43}{4}\beta \right) \sqrt{1-e_0^2} + \left(-\frac{55}{4} - \frac{127}{8}\gamma - \frac{3}{4}\beta \right) \right]$$
$$\times (k_0 \, \sin\lambda - h_0 \, \cos\lambda)$$
$$+ \left[\left(\frac{1 - \sqrt{1-e_0^2}}{e_0^2} \right) \left(-1 - 2\gamma - \frac{5}{2}\beta \right) + \left(\frac{11}{2} + 6\gamma - \frac{5}{4}\beta \right) \right]$$
$$\left. \times ((k_0^2 - h_0^2) \, \sin 2\lambda - 2h_0 k_0 \, \cos 2\lambda) + \cdots \right\} \qquad (4.4.1b)$$

$$\Delta k(t) = \frac{\mu}{a_0(1-e_0^2)} \left\{ -(2 + 2\gamma - \beta)h_0 n_0 t \right.$$
$$+ \left[k_0 h_0 \left(-8 - \frac{31}{4}\gamma + \frac{5}{2}\beta \right) \right] \sin\lambda$$
$$+ \left[(-\gamma - 2\beta) + \left(6 + \frac{63}{8}\gamma + \frac{3}{4}\beta \right) h_0^2 + \left(-2 + \frac{1}{8}\gamma + \frac{13}{4}\beta \right) k_0^2 \right] \cos\lambda$$
$$+ \left[\left(-1 - 2\gamma - \frac{5}{2}\beta \right) h_0 + \left(\frac{13}{2} + \frac{26}{3}\gamma + \frac{37}{12}\beta \right) h_0^3 \right]$$

---

† Here we restrict ourselves to terms of order $\mathcal{O}(e_0^2)$ with $\mathcal{O}(e_0) \simeq \mathcal{O}(k_0) \simeq \mathcal{O}(h_0)$. δa is a constant modifying the value of the "unperturbed" semi-major axis.

$$+ \left(-\frac{13}{2} - 4\gamma + \frac{25}{4}\beta \right) h_0 k_0^2 \Bigg] \sin 2\lambda$$

$$+ \left[\left(-1 - 2\gamma - \frac{5}{2}\beta \right) k_0 + \left(13 + 15\gamma + \frac{3}{2}\beta \right) k_0 h_0^2 \right.$$

$$\left. + \left(\frac{7}{3}\gamma + \frac{14}{3}\beta \right) k_0^3 \right] \cos 2\lambda + \cdots \Bigg\} \tag{4.4.1c}$$

$$\Delta h(t) = \frac{\mu}{a_0(1 - e_0^2)} \Bigg\{ (2 + 2\gamma - \beta) k_0 n_0 t$$

$$+ \left[(-\gamma - 2\beta) + \left(6 + \frac{63}{8}\gamma + \frac{3}{4}\beta \right) k_0^2 + \left(-2 + \frac{1}{8}\gamma + \frac{13}{4}\beta \right) h_0^2 \right] \sin \lambda$$

$$+ \left[k_0 h_0 \left(-8 - \frac{31}{4}\gamma + \frac{5}{2}\beta \right) + (k_0^2 + h_0^2)\left(29over6 + \frac{109}{24}\gamma - \frac{5}{12}\beta \right) \right] \cos \lambda$$

$$+ \left[\left(-1 - 2\gamma - \frac{5}{2}\beta \right) k_0 + \left(\frac{13}{2} + \frac{26}{3}\gamma + \frac{37}{12}\beta \right) k_0^3 \right.$$

$$\left. + \left(-\frac{13}{2} - 4\gamma + \frac{25}{4}\beta \right) h_0^2 k_0 \right] \sin 2\lambda$$

$$+ \left[\left(1 + 2\gamma + \frac{5}{2}\beta \right) h_0 - \left(13 + 15\gamma + \frac{3}{2}\beta \right) h_0 k_0^2 \right.$$

$$\left. - \left(\frac{7}{3}\gamma + \frac{14}{3}\beta \right) h_0^3 \right] \cos 2\lambda + \cdots \Bigg\} \,. \tag{4.4.1d}$$

For the Einstein case ($\gamma = \beta = 1$) one obtains (to first order in $GM_\odot/c^2$) the perturbations of planetary orbits in isotropic (or harmonic) coordinates that are indicated in Tabs. 4.5–4.8. (Lestrade et al. 1982):

Table 4.5 $10^{10}\Delta a$[A.U.]

	Const.	$\sin \lambda$	$\cos \lambda$	$\sin 2\lambda$	$\cos 2\lambda$
Mercury	−407.66	−284.54	−63.31	−34.04	+72.72
Venus	−394.84	−7.00	+6.21		
Earth-Moon	−394.91	−22.51	+5.17	+0.23	+0.47
Mars	−397.42	+52.64	−118.58	+12.17	−11.00
Jupiter	−395.90	−16.63	−65.00	−2.12	−3.88
Saturn	−395.85	−76.76	+4.10	+0.62	+5.75
Uranus	−395.48	−7.80	+63.58		
Neptune	−394.87	−9.25	−8.29		

Here the time T is indicated in Julian centuries, reckoned from the epoch J2000. The mean longitudes of the planets can be found in Table 3.1 (see also Lestrade et al. 1982).

Table 4.6 $10^{10}\Delta\lambda$[rad]

	$\sin\lambda$	$\cos\lambda$	$\sin 2\lambda$	$\cos 2\lambda$
Mercury	+110.76	−497.79	−73.86	−34.58
Venus	−5.82	−6.57		
Earth-Moon	−3.51	−15.27		
Mars	+52.78	+23.43	+2.85	+3.15
Jupiter	+8.49	−2.17		
Saturn	−0.29	−5.45		
Uranus	−2.24	−0.28		
Neptune	+0.19	−0.21		

Table 4.7 $10^{10}\Delta k$

	T	$\sin\lambda$	$\cos\lambda$	$\sin 2\lambda$	$\cos 2\lambda$
Mercury	−418 261.635	−30.73	−643.80	−256.13	−51.57
Venus	−2 118.752	+0.04	−409.35	−3.80	+3.37
Earth-Moon	−3 030.663	+0.08	−295.82	−8.84	+2.03
Mars	+2 482.256	+2.79	−194.02	+13.63	−30.16
Jupiter	−36.313	−0.14	−57.01	−1.26	−4.90
Saturn	−36.673	+0.02	−30.63	−3.13	+0.17
Uranus	−0.649	+0.02	−15.42	−0.16	+1.30
Neptune	−0.250	−0.00	−9.83		

Table 4.8 $10^{10}\Delta h$

	T	$\sin\lambda$	$\cos\lambda$	$\sin 2\lambda$	$\cos 2\lambda$
Mercury	+93 063.885	−775.06	−30.73	−66.80	+275.87
Venus	−1 878.571	−409.36	+0.04	+3.37	+3.80
Earth-Moon	−696.171	−296.14	+0.08	+2.03	+8.84
Mars	+5591.033	−188.98	+2.79	−29.98	−13.07
Jupiter	+142.134	−56.49	−0.14	−4.88	+1.24
Saturn	−1.958	−31.05	+0.02	+0.17	+3.15
Uranus	−5.292	−15.28	+0.02	+1.29	+0.16
Neptune	+0.224	−9.83	−0.00		

    Without doubt, the most famous relativistic effect in the motion of a planet is the anomalous secular perihelion shift $\Delta\omega$, that can be obtained from Tabs. 4.7 and 4.8 by:

$$\Delta\omega \simeq \left(1 + \frac{h_0^2}{k_0^2}\right)^{-1}\left(\frac{\Delta h}{k_0} - \frac{\Delta k}{k_0}\frac{h_0}{k_0}\right) \, . \tag{4.4.2}$$

For Mercury we obtain the well known value of $\Delta\omega \simeq 42.98''$ per century. For the Einstein case the relativistic secular perihelion precession is given in Table

4.9 for Mercury to Saturn. For Uranus and Pluto it is smaller than 0.01″ per century.

Table 4.9 Einsteinian Perihelion Precession of the Planets.

	Me	V	EM	Ma	J	S
$\Delta\omega[''/100\mathrm{y}]$	42.98	8.62	3.84	1.35	0.06	0.01

Furthermore, from Table 4.5 we see that the semi-major axes of the planetary orbits in spatially isotropic coordinates are subject to periodic perturbations with amplitudes 4.5 km (Me), 100 m (V), 300 m (EM), 750 m (M), 250 m (J), 1.2 km (S), 120 m (U) and 140 m (N), respectively.

The Lense-Thirring precession of Mercury due to the solar angular momentum is of order 0.01″ per century and those of the other planets are correspondingly smaller. Of interest are also the relativistic perihelion drifts for some moons in the planetary system. The Einstein perihelion precession can be as large as $\sim$ 270″/century for Jupiter's moon Io or $\sim$ 720″/century for Saturn's moon 1980S28. For these two moons the nodal drifts due to the Lense-Thirring precession amount to about 9 (Io) and 50 (1980S28) arcsec per century.

4.5 Numerically Produced Ephemerides

Due to the rapid progress in Computer techniques and suitable machine languages it has become convenient to produce high precision ephemerides for the Sun, Moon and planets on a purely numerical basis. Currently, the most widely used purely numerically produced ephemeris is DE200† of the *Jet Propulsion Laboratory* (JPL). Table 4.10 shows the structure of the program GLE2000‡ that is capable to reproduce the DE118(DE200) data with high accuracy. For example the agreement for the Earth-Moon distance is in the mm range for an integration time of 10 years. All initial values for positions, Euler angles and velocities as well for the dynamical parameters (masses, moments of inertia etc.) have been taken from the DE118 data. A complete listing of the GLE2000 FORTRAN code can be found in Schastok et al. (1988).

---

† DE200 refers to the epoch J2000 and differs from DE118 only by a constant rotation.

‡ GLE2000 was originally designed by GLEixner at the TU in Munich and was improved and extended in collaboration with the University in Tübingen.

Table 4.10   Structure of the Ephemeris Program GLE2000

GLE2000 (TU Munich, University of Tübingen)

(A) POINT DYNAMICS (EIH Equations)

SUN, MOON, all PLANETS incl. PLUTO
MP: CERES, PALLAS, VESTA, IRIS, BAMBERGA
SUN: $\mathbf{x}, \dot{\mathbf{x}}$ from PN-CENTER OF MASS + MOMENTUM CONDITION

(B) DYNAMICS OF EXTENDED BODIES (Newtonian)

EARTH: Y_2^0, Y_3^0, Y_4^0 $\rightarrow$ SUN
                                 $\rightarrow$ MOON

                                             TIDAL EFFECTS on
                                             translational and
                                             rotational motion

MOON: Y_2^0, Y_3^0, Y_4^0
              $\rightarrow$ EARTH, SUN
         $Y_2^1, \ldots, Y_4^4$

ROTATION   MOON: EULER Equation for rigid MOON
                    TORQUES: SUN, EARTH
            EARTH: PRECESSION, NUTATION from WAHR's model

(C) DISSIPATIVE EFFECTS (HEURISTICALLY)
TIDAL FRICTION: EARTH – MOON

CONSTANTS and INITIAL CONDITIONS (FROM DE118)

LIGHT VELOCITY c $R_\oplus$, R_C LOVE-NUMBER k_2
(A) MASSES   (B) C_{lm}, S_{lm} (C) PHASE ANGLE δ
    $\mathbf{x}_a(t_0)$, $\dot{\mathbf{x}}_a(t_0)$ EULER ANGLES
                       MOMENTS OF INERTIA

4.6 Timing Observations of Pulsars in Binary Systems

In this Section we will present one post-Newtonian model for the analysis of radio pulses from a pulsar in a binary system based on the Wagoner-Will representation (A.2 of the Appendix) of the solution to the PPN two-body problem. A similar treatment for the Einstein case was given by Haugan (1985). A different model, where also the effects of aberration are included, has been worked out by Damour et al. (1986).

Blandford et al. (1976) developed what has become the standard method for extracting information from the pulsar timing data. Their timing model is of lower accuracy only because the motion of the binary system was treated at Newtonian order. It was extended to include post-Newtonian corrections to the orbital motion by Epstein (1977) and Haugan (1985).

We chose standard post-Newtonian coordinates $(t, \mathbf{x})$ with origin in the "center of mass" of the binary system. A reference plane (Fig. 4.3) is defined to be a plane perpendicular to the line of sight from the Earth (more precisely the barycenter) to the pulsar ("plane of the sky") and a reference direction (defining the ascending node of the orbits) points from the origin to the north celestial pole.

The time of emission of the Nth radio pulse is given implicitly by

$$N = N_0 + \nu\tau + \frac{1}{2}\dot{\nu}\tau^2 + \frac{1}{6}\ddot{\nu}\tau^3 + \cdots , \qquad (4.6.1)$$

where $\dot{\nu} \equiv d\nu/d\tau|_{\tau=0}$ and higher derivatives will be ignored. Here, τ is up to a constant scale factor equal to the proper time of emission measured by a clock in an inertial frame at the surface of the pulsar, N_0 is an arbitrary constant and ν is the rotation frequency of the pulsar. According to (3.2.2) τ is related to the coordinate time of emission t_{em} by (1: pulsar; 2: companion):

$$\frac{d\tau}{dt_{\text{em}}} = 1 - U(\mathbf{x}_1) - \frac{1}{2}\mathbf{v}_1^2 + \mathcal{O}(\epsilon^4) . \qquad (4.6.2)$$

To sufficient accuracy ($m = m_1 + m_2$; $r = |\mathbf{x}_1 - \mathbf{x}_2|$):

$$\mathbf{v}_1^2 \simeq \frac{m_2^2}{m}\left(\frac{2}{r} - \frac{1}{a}\right) \quad ; \quad U(\mathbf{x}_1) \simeq \frac{Gm_2}{r}$$

and with

$$r \simeq a(1 - e\cos E) \quad ; \quad E - e\sin E \simeq t/\mathcal{P} + \sigma$$

(4.6.2) can be integrated to yield

$$\tau = t_{\text{em}} - \mathcal{C}\sin E(t_{\text{em}}) \qquad (4.6.3)$$

with ($a_1 \simeq am_2/m$):

$$\mathcal{C} \equiv \left(\frac{Gm_2}{a_1}\right)\left(\frac{m_2}{m}\right)\left(\frac{m_1 + 2m_2}{m}\right)\mathcal{P}e . \qquad (4.6.4)$$

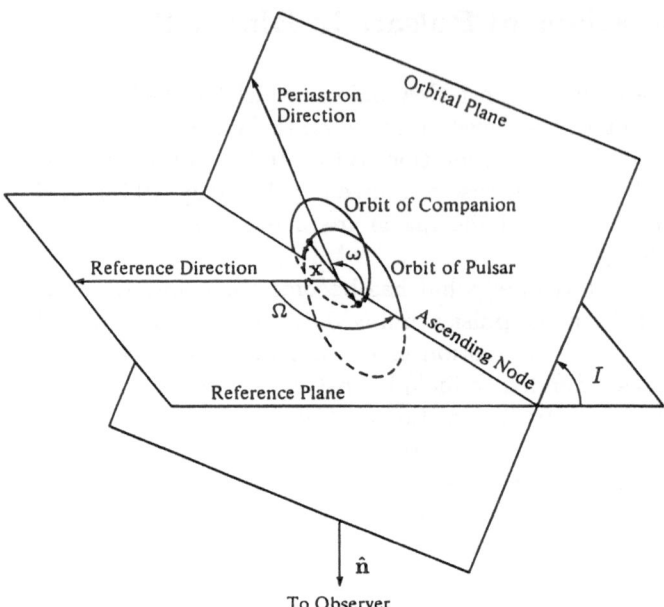

Fig. 4.3. Geometry for the binary pulsar system

Here, and in the following, we drop multiplicative and additive constants. Multiplicative constants can be absorbed in the definition of ν, additive constants in the value for N_0.

We first consider a distant observer at rest w.r.t. the coordinate system or the barycenter of the binary system. From (3.3.21) and (3.3.24a) we see that

$$t_{\text{arr}} - t_{\text{em}} = \int_{\mathbf{x}_1(t_{\text{em}})}^{\mathbf{x}_{\text{arr}}(t_{\text{arr}})} (1 + \epsilon + (\gamma + 1)U)\, dx_\parallel \;, \qquad (4.6.5)$$

where ϵ indicates the effect of the interstellar dispersion causing the pulses to travel with a group velocity less than unity as measured in the proper reference frame of the observer ($v_{\text{prop}} = (1-\epsilon)c$). The U-term describes the gravitational time delay discussed further in Section 5.3.1. Usually the integral over ϵ is written as D/f^2, where f is the frequency of the radiation and the *dispersion constant* D is given by (m_e: electron mass; DM: *dispersion measure*):

$$D = \frac{e^2}{2\pi m_e} \int n_e\, dx_\parallel \equiv \frac{e^2}{2\pi m_e}\, DM \;. \qquad (4.6.6)$$

The integral in (4.6.5) can be performed if we consider the binary system to be stationary during the pulse emission. Again dropping constant terms we obtain

$$\int_{\mathbf{x}_1}^{\mathbf{x}_{\text{arr}}} U\, dx_\parallel \simeq Gm_2 \int_{t_{\text{em}}}^{t_{\text{arr}}} \frac{dt}{|\mathbf{x}(t) - \mathbf{x}_2(t_{\text{em}})|} \;,$$

where the integration is taken over the unperturbed orbit. With

$$\mathbf{x}(t) = \mathbf{x}_1(t_{\mathrm{em}}) + [\mathbf{x}_{\mathrm{arr}}(t_{\mathrm{arr}}) - \mathbf{x}_1(t_{\mathrm{em}})] \, \frac{t - t_{\mathrm{em}}}{t_{\mathrm{arr}} - t_{\mathrm{em}}}$$

the integral gives ($\mathbf{x} = \mathbf{x}_{\mathrm{arr}}; \mathbf{r} = \mathbf{x}_1 - \mathbf{x}_2$):

$$Gm_2 \, \frac{t_{\mathrm{arr}} - t_{\mathrm{em}}}{|\mathbf{x}_{\mathrm{arr}} - \mathbf{x}_1|} \, \ln \left[\frac{|\mathbf{x} - \mathbf{x}_1||r + (\mathbf{x} - \mathbf{x}_1)| + |\mathbf{x} - \mathbf{x}_1|^2 + \mathbf{r} \cdot (\mathbf{x} - \mathbf{x}_1)}{|\mathbf{x} - \mathbf{x}_1|r + \mathbf{r} \cdot (\mathbf{x} - \mathbf{x}_1)} \right]$$

$$\simeq Gm_2 \, \ln \left(\frac{2x}{r + \mathbf{r} \cdot \hat{\mathbf{n}}} \right) \qquad \text{for} \quad x = |\mathbf{x}| \gg |\mathbf{x}_1| \ , \tag{4.6.7}$$

where $\hat{\mathbf{n}} = \mathbf{x}_{\mathrm{arr}}/x$ (Fig. 4.3) points from the origin to the observer. Hence,

$$t_{\mathrm{arr}} - t_{\mathrm{em}} = |\mathbf{x}_{\mathrm{arr}}(t_{\mathrm{arr}}) - \mathbf{x}_1(t_{\mathrm{em}})| + \frac{D}{f^2} + (\gamma + 1)Gm_2 \, \ln \left[\frac{2x}{r + \mathbf{r} \cdot \hat{\mathbf{n}}} \right]_{\mathrm{em}} \ . \tag{4.6.8}$$

We now write

$$\mathbf{x}_{\mathrm{arr}}(t_{\mathrm{arr}}) = \mathbf{x}_b(t_{\mathrm{arr}}) + \mathbf{x}_{bs}(t_{\mathrm{arr}}) \ , \tag{4.6.9}$$

where b refers to the barycenter of the solar system and s to the observing station at the Earth, and get

$$|\mathbf{x}_{\mathrm{arr}}(t_{\mathrm{arr}}) - \mathbf{x}_1(t_{\mathrm{em}})| \simeq r_b(t_{\mathrm{arr}}) + \mathbf{x}_{bs}(t_{\mathrm{arr}}) \cdot \hat{\mathbf{n}} - \mathbf{x}_1(t_{\mathrm{em}}) \cdot \hat{\mathbf{n}} \ . \tag{4.6.10}$$

Therefore,

$$t_{\mathrm{arr}} = t_{\mathrm{em}} + r_b(t_{\mathrm{arr}}) + \mathbf{x}_{bs}(t_{\mathrm{arr}}) \cdot \hat{\mathbf{n}} - \mathbf{x}_1(t_{\mathrm{em}}) \cdot \hat{\mathbf{n}}$$

$$+ \frac{D}{f^2} + (\gamma + 1)Gm_2 \, \ln \left[\frac{2x}{r + \mathbf{r} \cdot \hat{\mathbf{n}}} \right]_{\mathrm{em}} \ . \tag{4.6.11}$$

Definition.

$$t = t_{\mathrm{arr}} - r_b(t_{\mathrm{arr}}) - \mathbf{x}_{bs}(t_{\mathrm{arr}}) \cdot \hat{\mathbf{n}} - \frac{D}{f^2} \tag{4.6.12}$$

is called the infinite frequency barycentric arrival time ("arrival time"), where f is measured from the barycentric system ($f \simeq f_s(1 - \mathbf{v}_s \cdot \hat{\mathbf{n}})$).

In the following the arrival time t defined in (4.6.12) will be related to the pulsar proper emission time τ, assuming that the observed value of arrival time measured in proper time of the station clock has been transformed to barycentric coordinate time according to our discussion from Section 3.2. From (4.6.11) we have

$$t = t_{\mathrm{em}} - \mathbf{x}_1(t_{\mathrm{em}}) \cdot \hat{\mathbf{n}} + (\gamma + 1)Gm_2 \, \ln \left[\frac{2x}{r + \mathbf{r} \cdot \hat{\mathbf{n}}} \right]_{\mathrm{em}} \ , \tag{4.6.13}$$

where the second term describes the integrated effect of first order Doppler shift and the third term the gravitational time delay in the field of the companion. We will now analyze the last two terms in (4.6.13) by using the Wagoner-Will representation of the post-Newtonian orbit described in A.2. $\mathbf{x}_1 \cdot \hat{\mathbf{n}}$ can be written as

$$\mathbf{x}_1 \cdot \hat{\mathbf{n}} = -r_1 \sin I \sin \phi \;, \tag{4.6.14}$$

where ϕ is reckoned from the line of nodes. According to (A2.18) we will write

$$\phi = \left[1 + (2\gamma + 2 - \beta)\frac{m}{p}\right]\dot{\eta} + \omega_0 \tag{4.6.15}$$

and r_1 can be obtained from (A2.5a):

$$r_1 = \left[\frac{m_2}{m} + \frac{\mu\delta m}{2m^2}\left(\mathbf{v}^2 - \frac{Gm}{r}\right)\right] r \tag{4.6.16}$$

with $\delta m = m_1 - m_2$ and $\mu = m_1 m_2/m$. With

$$\mathbf{v}^2 - \frac{Gm}{r} = Gm\left(\frac{1}{r} - \frac{1}{a}\right) + \mathcal{O}(\epsilon^2)$$

we can obtain an expression for r_1 using (A2.23) to express r as function of the eccentric anomaly E that is related to the coordinate time by means of the generalized Kepler equation (A2.26). The result reads:

$$r_1 = a_1 \left(1 - e' \cos E - e'' \cos 2E\right) \tag{4.6.17}$$

with ($\nu \equiv \mu/m$)

$$a_1 = \frac{m_2 a}{m}\left[1 + \frac{m}{4a(1 - e^2)^2}\left\{4(2\beta + \gamma) + (10\gamma + \beta + 12)e^2 - 2\gamma e^4\right.\right.$$
$$\left.\left. - (4 + 17e^2 + 3e^4)\nu\right\}\right]$$

$$e' = e\left[1 + \frac{m}{4a(1 - e^2)^2}\{(12\gamma + 6\beta + 8) - (10\gamma + 3\beta + 4)e^2 + 2\gamma e^4\right.$$
$$\left. - (8 + 3e^2 - 3e^4)\nu\} - \frac{m_1\delta m}{2ma}\right] \equiv e + \Delta e$$

and

$$e'' = \frac{e^2 m}{4a(1 - e^2)^2}\left[-(6\gamma + 3\beta + 4) + 2\gamma e^2 + (3 + 5e^2)\nu\right] \;.$$

In the argument of the time delay term,

$$(r + \mathbf{r} \cdot \hat{\mathbf{n}})^{-1} \simeq p^{-1}\frac{1 + e \cos \eta}{1 - \sin I \sin(\eta + \omega)} \;, \tag{4.6.18}$$

where for the remaining part of this section

$$\omega = \omega_{\text{rel}} = \omega_0 + \dot{\omega}_{\text{PN}}\,(t - T_0)\ . \tag{4.6.19}$$

Here $\dot{\omega}_{\text{PN}}$ denotes the post-Newtonian value for the relativistic perihelion precession given by (4.2.15):

$$\dot{\omega}_{\text{PN}} = \frac{(2\gamma + 2 - \beta)Gm}{c^2 a(1 - e^2)}\,n_0\ . \tag{4.6.20}$$

Equation (4.6.13) can therefore be written as (dropping constant terms):

$$\begin{aligned}
t_{\text{em}} = t &- a_1 \sin I(1 - e' \cos E - e'' \cos 2E) \sin\left[\left(1 + (2\gamma + 2 - \beta)\frac{m}{p}\right)\eta + \omega_0\right] \\
&- (\gamma + 1)Gm_2\,\ln\left[\frac{1 + e\cos\eta}{1 - \sin I \sin(\eta + \omega)}\right]\ . \tag{4.6.21}
\end{aligned}$$

Since

$$\begin{aligned}
\eta + (2\gamma + 2 - \beta)\frac{m}{p}\eta + \omega_0 &= \eta + (2\gamma + 2 - \beta)\frac{m}{p}(\eta - n_0(t - T_0)) \\
&\quad + \omega_0 + (2\gamma + 2 - \beta)\frac{m}{p}n_0(t - T_0) \\
&= \eta + (2\gamma + 2 - \beta)\frac{m}{p}(\eta - M) + \omega \tag{4.6.22}
\end{aligned}$$

this result can also be written in the form

$$\begin{aligned}
t_{\text{em}} = t &- a_1 \sin I\,(1 - e' \cos E - e'' \cos 2E)\sin(\eta + \omega) \\
&- a_1 \sin I\,(1 - e \cos E)\cos(\eta + \omega)\left[(2\gamma + 2 - \beta)\frac{m}{p}(\eta - M)\right] \\
&- (\gamma + 1)Gm_2\,\ln\left[\frac{1 + e\cos\eta}{1 - \sin I \sin(\eta + \omega)}\right]\ . \tag{4.6.23}
\end{aligned}$$

Note that this equation does not yet express t_{em} as a function of t explicitly since $E = E(t_{\text{em}})$. Since $t_{\text{em}} - t$ is of order $\mathcal{O}(\epsilon)$ we can solve for $t_{\text{em}} = t_{\text{em}}(t)$ by iteration formally to $\mathcal{O}(\epsilon^3)$. We first notice that the Δe and e'' terms in the first line and the terms from the second and third line in (4.6.23) are already of order $\mathcal{O}(\epsilon^3)$. With

$$\begin{aligned}
\sin(\eta + \omega) &= (1 - e\cos E)^{-1} \\
&\quad \times [(\cos E - e)\sin\omega + (1 - e^2)^{1/2}\sin E \cos\omega] + \mathcal{O}(\epsilon) \\
\cos(\eta + \omega) &= (1 - e\cos E)^{-1} \\
&\quad \times [(\cos E - e)\cos\omega - (1 - e^2)^{1/2}\sin E \sin\omega] + \mathcal{O}(\epsilon)\ ,
\end{aligned}$$

where $E = E(t)$, here and in the following, these ϵ^3 terms together yield

$$\frac{\Delta t}{1 - e \cos E}[\Delta e \cos E + e'' \cos 2E]$$
$$- (2\gamma + 2 - \beta)\left(\frac{m}{p}\right)(\eta - M)a_1 \sin I$$
$$\times [(\cos E - e) \cos \omega - (1 - e^2)^{1/2} \sin E \sin \omega]$$
$$+ (\gamma + 1)Gm_2 \ln\{1 - e \cos E - \sin I$$
$$\times [(1 - e^2)^{1/2} \sin E \cos \omega + (\cos E - e) \sin \omega]\} \qquad (4.6.24)$$

with

$$\Delta t \equiv \mathcal{A}(\cos E - e) + \mathcal{B} \sin E , \qquad (4.6.25)$$

where

$$\mathcal{A} \equiv \alpha \sin \omega$$
$$\mathcal{B} \equiv (1 - e^2)^{1/2}\alpha \cos \omega$$

and

$$\alpha \equiv a_1 \sin I .$$

Hence, only the remaining part

$$t_{\text{em}} - t = -[\mathcal{A}(\cos E(t_{\text{em}}) - e) + \mathcal{B} \sin E(t_{\text{em}})] \qquad (4.6.26)$$

has to be iterated. In a first step we may take

$$E(t_{\text{em}}) = E(t) + \mathcal{O}(\epsilon) = E + \mathcal{O}(\epsilon)$$

and find

$$t_{\text{em}} - t = -\Delta t + \mathcal{O}(\epsilon^2) . \qquad (4.6.27)$$

In the next iteration we start with

$$E(t_{\text{em}}) = E + \dot{E}|_t (t_{\text{em}} - t) + \mathcal{O}(\epsilon^2) = E - \frac{\mathcal{P}^{-1}\Delta t}{1 - e \cos E} + \mathcal{O}(\epsilon^2) ,$$

where $\dot{E}$ was evaluated from the (Newtonian) Kepler equation. Therefore,

$$t_{\text{em}} - t = -\Delta t + \frac{\Delta t}{1 - e \cos E}\left(\frac{\mathcal{B}}{\mathcal{P}}\cos E - \frac{\mathcal{A}}{\mathcal{P}}\sin E\right) + \mathcal{O}(\epsilon^3) . \qquad (4.6.28)$$

In the last iteration we take

$$E(t_{\text{em}}) = e + \dot{E}(t_{\text{em}} - t) + \frac{1}{2}\ddot{E}(t_{\text{em}} - t)^2 + \mathcal{O}(\epsilon^3) \equiv E + \Delta E + \mathcal{O}(\epsilon^3) ,$$

where $(t_{em} - t)$ now has to be taken into account to $\mathcal{O}(\epsilon^2)$. Using (4.6.28) and

$$\dot{E} \simeq \frac{\mathcal{P}^{-1}}{1 - e\cos E} \quad ; \quad \ddot{E} \simeq -\frac{\mathcal{P}^{-2}e\sin E}{(1 - e\cos E)^3}$$

ΔE is finally found to be

$$\Delta E = \frac{\mathcal{P}^{-1}}{1 - e\cos E}\left[-\Delta t + \frac{\Delta t}{1 - e\cos E}\left(\frac{\mathcal{B}}{\mathcal{P}}\cos E - \frac{\mathcal{A}}{\mathcal{P}}\sin E\right)\right]$$
$$-\frac{\mathcal{P}^{-2}e\sin E}{2(1 - e\cos E)^3}\Delta t^2 \ . \tag{4.6.29}$$

The final iteration of (4.6.26) is then performed with

$$\cos E(t_{em}) = \cos E - \Delta E \ \sin E - \tfrac{1}{2}(\Delta E)^2\cos E + \mathcal{O}(\epsilon^3)$$

$$\sin E(t_{em}) = \sin E + \Delta E \ \cos E - \tfrac{1}{2}(\Delta E)^2\sin E + \mathcal{O}(\epsilon^3) \ .$$

Adding the terms from (4.6.24) and converting t_{em} to τ using (4.6.3) we finally get:

$$\begin{aligned}
\tau = {}& t - \mathcal{A}(\cos E - e) - (\mathcal{B} + \mathcal{C})\sin E \\
& - (2\gamma + 2 - \beta)\left(\frac{m}{p}\right)(\eta - M)a_1\sin I \\
& \quad \times [(\cos E - e)\cos\omega - (1 - e^2)^{1/2}\sin E\sin\omega] \\
& + (\gamma + 1)Gm_2\ \ln[1 - e\cos E - \sin I \\
& \quad \times \{(1 - e^2)^{1/2}\sin E\cos\omega + (\cos E - e)\sin\omega\}] \\
& + \frac{\Delta t}{1 - e\cos E}\left[\left(\frac{\mathcal{B} + \mathcal{C}}{\mathcal{P}} + \Delta e\right)\cos E - \frac{\mathcal{A}}{\mathcal{P}}\sin E + e''\cos 2E\right. \\
& + \frac{1}{2\mathcal{P}^2(1 - e\cos E)}\ [(\mathcal{A}\cos E + \mathcal{B}\sin E)\Delta t \\
& + (\mathcal{B}\cos E - \mathcal{A}\sin E)\left\{2(\mathcal{A}\sin E - \mathcal{B}\cos E)\right. \\
& \left.\left.+ \frac{e\sin E}{1 - e\cos E}\Delta t\right\}\right] \ , \tag{4.6.30}
\end{aligned}$$

where $\mathcal{A}, \mathcal{B}, \mathcal{C}, \Delta t, \Delta e$ and e'' are given by (4.6.4, 4.6.17 and 4.6.25). Together with (4.6.1) and the Kepler equation (A2.26) this provides the final timing-model equation

$$N = N(t; N_0, \nu, \dot{\nu}, \ddot{\nu}, \alpha, \omega, e, \sin I, \mathcal{P}, \sigma, \mathcal{C}) \ . \tag{4.6.31}$$

For various reasons the motion of a binary system like e.g. PSR 1913+16 will not precisely follow the post-Newtonian two-body point-mass orbit assumed in this model. Reasons for this might be: tidal or rotational distortion of the companion, dissipation of orbital energy either by viscous processes or by gravitational radiation, mass loss by the system etc. (see e.g. Smarr et al. 1976, Will 1981, Straumann 1984). These perturbing effects might be taken into account by *osculating post-Newtonian parameters*. Secular changes of α, ω, e etc. can be included in the timing formula (4.6.31) by the replacements

$$\alpha \to \alpha + \dot{\alpha}t \quad ; \quad \omega + \dot{\omega}t$$

$$e \to e + \dot{e}t \quad ; \quad \mathcal{P} \to \mathcal{P} + \tfrac{1}{2}\dot{\mathcal{P}}t,$$

where the factor $1/2$ in front of $\dot{\mathcal{P}}$ comes from (Blandford et al. 1976):

$$E - e \sin E = m^{1/2} \int a^{-3/2}\, dt + \sigma$$

$$\simeq m^{1/2} \int (a + \dot{a}t)^{-3/2}\, dt + \sigma$$

$$\simeq \frac{t}{\mathcal{P} + \tfrac{1}{2}\dot{\mathcal{P}}t} + \sigma$$

with

$$\dot{\mathcal{P}} = \frac{3}{2}\left(\frac{a^3}{m}\right)^{1/2}\left(\frac{\dot{a}}{a}\right)\ .$$

A secular variation of σ merely modifies the apparent unperturbed value of $\mathcal{P}$.

4.7 Relativistic Motion of Extended Bodies

Seventy years now have passed since Einstein's paper on the foundations of General Relativity appeared in the literature and yet there is no satisfactory and practicable description of the motion of extended bodies in Einstein's theory of gravity.

The full problem of the dynamics of extended bodies in Einstein's theory of gravity is summarized in Fig. 4.4, which arose out of a discussion with Jürgen Ehlers. The starting points of further theoretical considerations are: Einstein's field equations for the metric tensor g (I), equations of state specifying a model of matter (II) and a certain set of boundary constraints such as the requirements of asymptotic flatness for g or no incoming or outgoing gravitational radiation (III). If the model of matter is compatible with the field equations all information about the dynamics of matter is embodied in the local equations of motion as given by the divergencelessness of the energy momentum tensor

$$\nabla_\mu T^{\mu\nu} = 0\ . \tag{4.7.1}$$

Since the metric tensor g appears explicitly in (4.7.1), it is called an equation of motion of the first kind. It is obvious that such an equation is not very useful for the description of the motion of bodies in the solar system. In the Newtonian framework one successfully proceeds with the introduction of collective variables such as centres of mass, mass multipole moments, angular and translational momenta etc. and derives global equations of motion for the momentum and angular momentum vector where the Newtonian potential can be expressed in terms of the multipole moments of the sources. Such an equation of motion is said to be of the second kind. Dixon (1970) has characterized the steps that are necessary to generalize this Newtonian route to the full Einstein theory:

    i) suitable choice of a representative point (center of mass) within each body
   ii) derivation of a momentum-velocity relation
  iii) evaluation of the total force and torque exerted on an extended body in a gravitational field
  iv) characterization of the self-field
   v) evaluation of the self-force and self-torque
  vi) determination of the external field in terms of matter variables (multipole moments) of the field-generating bodies

Whereas most of the steps (especially ii) are conceptually trivial in Newtonian space-time they are highly problematic in a relativistically curved space-time; the separation of the total field into an external and a self-part (iv) and especially the determination of the gravitational field in terms of matter variables of the sources seems to be almost hopeless at present. This situation is characterized by question marks in Fig. 4.4. In a series of papers Dixon (1970a, b, 1973, 1974, 1976) (see also Ehlers et al. 1977, Bailey et al. 1980, Schattner et al. 1984) has solved the first three parts i) – iii) of the whole problem in an exact way (i.e. without resorting to approximation schemes) and in terms of geometrical (i.e. coordinate independent) quantities. However, the complexity in the *construction* of dynamical quantities such as 4-momentum, angular momentum tensor and reduced mass multipole moments is so enormous that so far no application of Dixon's theory exists in the literature even for simple external fields and it is not very likely that much progress will be achieved here in the near future.

    The other route, which is usually chosen in the literature, selects a certain set of coordinates from the beginning, usually by imposing a certain gauge condition. That implies that the covariance of the theory is broken at a very early stage. Once the coordinates are chosen the formulation of some approximation scheme, like e.g. that given by the post-Newtonian theory, usually is not very difficult. Here, the components of the metric tensor can then be expressed in terms of functionals of matter variables as is the case for our post-Newtonian metric (A1.1). If these functionals are inserted into the equations of motion (4.7.1) one obtains the local equations of motion of the second kind. One can then proceed to introduce a set of collective variables such as e.g. momentum

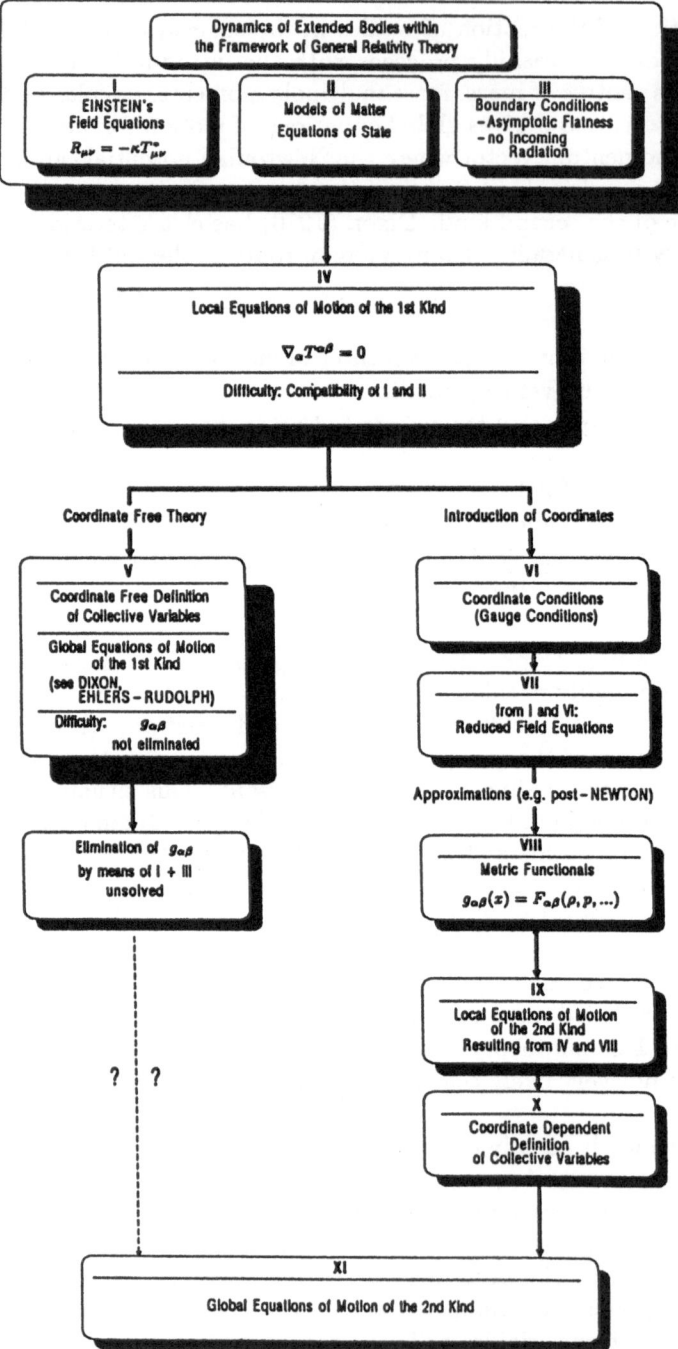

Fig. 4.4. Problem of dynamics of extended bodies in Einstein's theory of gravity

and angular momentum vector *according to the chosen coordinates* (cf. Section 2.2) and use the local equations of motion to derive the desired *global* equations of motion for the centers of mass, angular momenta etc. In contrast to the remaining part of this book, in this section and in A4 we will not use the standard post-Newtonian gauge but the so-called *harmonic gauge*†. The definition of *harmonic* coordinates reads:

$$\mathbf{g}^{\lambda\kappa}_{,\kappa} \equiv (\sqrt{-g}g^{\lambda\kappa})_{,\kappa} = 0 \ . \tag{4.7.2}$$

From the standard post-Newtonian coordinates the harmonic coordinates on the first PN level are obtained with a transformation of the time variable according to

$$t \longrightarrow t' = t - \frac{1}{2}\chi_{,0} \ , \tag{4.7.3}$$

where χ is the Newtonian superpotential (4.1.4). Hence, in harmonic coordinates the PPN metric reads:

$$g_{00} = -1 + 2U - 2\beta U^2 + 2\Phi - \chi_{,00} \tag{4.7.4a}$$
$$g_{0i} = -2(\gamma + 1)V_i \tag{4.7.4b}$$
$$g_{ij} = (1 + 2\gamma U)\delta_{ij} \ , \tag{4.7.4c}$$

where we have made use of

$$\chi_{,0i} = V_i - W_i \ .$$

Φ can be written as

$$\Phi = G \int \frac{\rho'\varphi'}{|\mathbf{x} - \mathbf{x}'|} \ d^3x' \tag{4.7.5}$$

with

$$\varphi = (\gamma + 1)\mathbf{v}^2 + (3\gamma - 2\beta + 1)U + \Pi + 3\gamma p/\rho \ . \tag{4.7.6}$$

Note that the transformation (4.7.3) leaves the Einstein-Infeld-Hoffmann equations of motion invariant, i.e. differences between harmonic and ADM gauge in the equations of motion will not show up before the PNA2 approximation.

Usually in the post-Newtonian framework matter is described as an ideal fluid for which the stress energy tensor takes the form

$$T^{\mu\nu} = (\rho(1 + \Pi) + p)u^\mu u^\nu + pg^{\mu\nu} \ , \tag{4.7.7}$$

---

† As in the case of the harmonic gauge it is possible to define coordinate systems in the full Einstein theory which reduce to the standard PN gauge in the *first* post-Newtonian approximation (PNA1). In the framework of the full Einstein theory this choice is usually called ADM gauge (Arnowitt et al. 1960). The ADM gauge has proven very useful both at the PNA2 (Ohta et al. 1974, Damour et al. 1987, 1988) and the PNA2.5 approximation level (Schäfer 1985).

where $u^\mu = dx^\mu/d\tau$ is the 4-velocity of the fluid, i.e. one neglects the anisotropic part of the stress tensor (for an exception see Misner et al. 1973). This, however, does not exclude the possibility of considering the bodies of an isolated system as *quasi-rigid*; the concept of quasi-rigidity can e.g. be achieved be considering suitably defined mass multipole moments of the bodies as practically independent of time.

Since

$$u^\mu \simeq \left(1 + \frac{1}{2}\mathbf{v}^2 + U, v^i u^0\right) \tag{4.7.8}$$

the necessary components of the energy momentum tensor read to PN order:

$$T^{00} = \rho(1 + \mathbf{v}^2 + 2U + \Pi) \tag{4.7.9a}$$
$$T^{0i} = \rho v^i(1 + \mathbf{v}^2 + 2U + \Pi + p/\rho) \tag{4.7.9b}$$
$$T^{ij} = \rho v^i v^j(1 + \mathbf{v}^2 + 2U + \Pi + p/\rho) + p\,\delta_{ij}(1 - 2\gamma U) \ . \tag{4.7.9c}$$

The local equations of motion are then given by the continuity equation

$$(\rho u^\mu)_{;\mu} = 0 \tag{4.7.10}$$

expressing the law of rest mass conservation, and the Euler equation of hydrodynamics

$$\left(T^{\mu i}\right)_{;\mu} = 0 \ . \tag{4.7.11}$$

In our coordinate system the continuity equation can also be written in the form

$$\frac{\partial \rho^*}{\partial t} + \nabla \cdot (\rho^* \mathbf{v}) = 0 \ , \tag{4.7.12}$$

where

$$\rho^* = \rho(-g)^{1/2} u^0 = \rho\left(1 + \frac{1}{2}\mathbf{v}^2 + 3\gamma U\right) \tag{4.7.13}$$

is the *conserved* post-Newtonian density. The Euler-equation can now be written in the following "quasi-Newtonian" form:

$$\frac{\partial}{\partial t}\pi^i + \frac{\partial}{\partial x^j}\sigma^{ij} = f^i \tag{4.7.14}$$

with

$$\pi^i = \rho^* v^i \left(1 + \frac{1}{2}\mathbf{v}^2 - U + \Pi + p/\rho^*\right) + (2\gamma + 2)\rho^*(v^i U - V_i) + \frac{1}{2}\rho^*(V_i - W_i) \tag{4.7.15}$$

$$\sigma^{ij} = \pi^i v^j + (1 + (3\gamma - 1)U)p\,\delta_{ij} \tag{4.7.16}$$

and

$$f^i = G\left[\rho U_{,i} + \rho^*(\varphi U_{,i} + \Phi_{,i}) - (2\gamma + 2)\rho^* v^j V_{j,i} + \frac{1}{2}\rho^* v^j \partial_j(V^i - W^i)\right] \ . \tag{4.7.17}$$

This form of the PPN-Euler equation has been derived by Chandrasekhar et al. (1969) for the Einstein case and by Caporali (1979) for the PPN-case; a derivation can be found in A.4. It is important to note that the momentum density π^i, the stress-density σ^{ij} and the force density f^i do not contain time derivatives so that (4.7.14) is a local equation of motion in a precise sense.

4.7.1 Translational Motion

Let us first concentrate upon the translational motion. If we now integrate (4.7.14) over the volume of a body a of an isolated N-body system we formally obtain a global equation of motion in the form

$$\frac{d}{dt}p^i = F^i \ , \tag{4.7.18}$$

where

$$p^i = \int_a \pi^i \, d^3x \tag{4.7.19}$$

and

$$F^i = \int_a f^i \, d^3x \ . \tag{4.7.20}$$

From (4.7.12) we see that the *rest mass* m_a^0 of body a, given by

$$m_a^0 = \int_a \rho^* d^3x \tag{4.7.21}$$

is a conserved quantity to PN order. Another conserved quantity is the so-called ADM mass (Arnowitt et al. 1960, Misner et al. 1973) of the total system:

$$m^{\mathrm{ADM}} = \int \Theta^{00} \, d^3x = \int \rho^* \left(1 + \frac{1}{2}\mathbf{v}^2 - \frac{1}{2}U + \Pi \right) d^3x \ , \tag{4.7.22}$$

where

$$\Theta^{\mu\nu} = (-g)(T^{\mu\nu} + t^{\mu\nu}) \tag{4.7.23}$$

is the *stress energy complex* and $t^{\mu\nu}$ are the components of the Landau-Lifshitz pseudo-tensor (Landau et al. 1962, Misner et al. 1973).† The stress-energy complex $\Theta^{\mu\nu}$ is defined such that

$$\Theta^{\mu\nu}{}_{,\nu} = 0 \tag{4.7.24}$$

---

† The components of the Landau-Lifshitz pseudo-tensor are given by $16\pi G(-g)t^{\alpha\beta} = g^{\alpha\beta}{}_{,\lambda}g^{\lambda\mu}{}_{,\mu} - g^{\alpha\lambda}{}_{,\lambda}g^{\beta\mu}{}_{,\mu} + \frac{1}{2}g^{\alpha\beta}g_{\lambda\mu}g^{\lambda\nu}{}_{,\rho}g^{\rho\mu}{}_{,\nu} - (g^{\alpha\lambda}g_{\mu\nu}g^{\beta\nu}{}_{,\rho}g^{\mu\rho}{}_{,\lambda} + g^{\beta\lambda}g_{\mu\nu}g^{\alpha\nu}{}_{,\rho}g^{\mu\rho}{}_{,\lambda}) + g_{\lambda\mu}g^{\nu\rho}g^{\alpha\lambda}{}_{,\nu}g^{\beta\mu}{}_{,\rho} + \frac{1}{8}(2g^{\alpha\lambda}g^{\beta\mu} - g^{\alpha\beta}g^{\lambda\mu})(2g_{\nu\rho}g_{\sigma\tau} - g_{\rho\sigma}g_{\nu\tau})g^{\nu\tau}{}_{,\lambda}g^{\rho\sigma}{}_{,\mu}$.

instead of $T^{\mu\nu}{}_{;\nu} = 0$, allowing the direct construction of conserved quantities. This suggests interpreting

$$E_a = \int_a \rho^* \left(\frac{1}{2} \bar{v}^2 - \frac{1}{2} U_{(a)} + \Pi \right) d^3x \tag{4.7.25}$$

as internal energy of the body a, if $\bar{v}$ is the *internal* velocity of the fluid w.r.t. some suitably chosen point in a.† Adding this internal energy E_a to the rest mass m_a^0 yields an expression for the total *inertial mass* m_a of body a:

$$m_a = \int_a \rho^* \left(1 + \frac{1}{2} \bar{v}^2 - \frac{1}{2} U_{(a)} + \Pi \right) d^3x \ . \tag{4.7.26}$$

$\bar{v}$ can be interpreted as

$$\bar{v} = v - v_{a(0)}$$

with

$$v_{a(0)} = m_a^{-1} \int_a \rho^* v \, d^3x \ .$$

The equation of motion for m_a can be derived from the continuity equation (4.7.12) and the energy momentum conservation

$$\Theta^{0\nu}{}_{,\nu} = 0 \ . \tag{4.7.27}$$

and using the Newtonian center of mass condition. One finds (Caporali 1979):

$$\frac{dm_a}{dt} = \frac{dE_a}{dt} = \int_a \rho \bar{v}^i \partial_i U_{(e)} \, d^3x \ , \tag{4.7.28}$$

where $U_{(e)}$ is the external potential produced by all bodies of the system other than a. This result simply indicates that the change in internal energy of the body a is given by the action of the external force on the fluid elements of body a. It is convenient to analyze the last relation by means of a multipole expansion of $U_{(e)}$ w.r.t. the Newtonian center of mass:

$$U_{(e)}(t, \mathbf{x}) = G \sum_{b \neq a} \left[\sum_{n=0}^{\infty} \frac{(-1)^n}{n!} m_{(b)}^{i_1 \dots i_n} \frac{\partial^n}{\partial x^{i_1} \dots \partial x^{i_n}} \left(\frac{1}{r_b} \right) \right] \tag{4.7.29}$$

with

$$m_{(b)}^{i_1 \dots i_n} = \int_b r_{(b)}^{i_1} \dots r_{(b)}^{i_n} \rho \, d^3x \ . \tag{4.7.30}$$

Here

$$\mathbf{r}_{(b)} = \mathbf{x} - \mathbf{x}_{(b)}$$

---

† The index (a) indicates that the integral defining the corresponding quantity should be taken over the volume of body a. $U_{(a)}$ is the self-gravitational potential of body a.

denotes the Newtonian "relative position vector" w.r.t. the Newtonian center of mass $\mathbf{x}_{(b)}$. Using (Dixon 1976)

$$p_{(a)}^{(i_1\ldots i_n)} = \frac{1}{n}\frac{d}{dt}m_{(a)}^{i_1\ldots i_n} \; , \tag{4.7.31}$$

where

$$p_{(a)}^{ji_1\ldots i_n} = \int_a \rho \bar{v}^j r_{(a)}^{i_1} \ldots r_{(a)}^{i_n} \, d^3x \; , \tag{4.7.32}$$

(4.7.28) can be written as:

$$\frac{dm_a}{dt} = \sum_{n=1}^{\infty} \frac{1}{n!}\frac{1}{(n+1)} \left(\frac{d}{dt}m_{(a)}^{ji_1\ldots i_n}\right) \left[\frac{\partial^{n+1}}{\partial x^{i_1}\ldots\partial x^{i_n}\partial x^j} U_{(e)}(t,\mathbf{x})\right]_{\mathbf{x}_a} . \tag{4.7.33}$$

For most astrophysical bodies the time evolution of m_a is governed by the time derivative of the quadrupole moment. For a rigidly rotating and spheroidal body (in the Newtonian sense; $m_{11} = m_{22} \neq m_{33}$ w.r.t. principal axes of inertia) the action of body b leads to

$$\frac{dm_a}{dt} \sim \frac{Gm_b}{c^2 r_{ab}^3}(m_{11} - m_{33})\,\omega_{1,2} \; , \tag{4.7.34}$$

i.e. a *tidal action* proportional to the difference of the principal moments of inertia and the components of angular velocity perpendicular to the symmetry axis of body a. For most astrophysical applications like the motion of the planetary system the right hand side of (4.7.34) or (4.7.28) is smaller than typical post-post-Newtonian (PNA2) terms and can be neglected to post-Newtonian order.

The conservation laws $\Theta^{\mu\nu}{}_{,\nu} = 0$ imply that the point defined by

$$\mathbf{X} = (m^{\mathrm{ADM}})^{-1} \int \rho^* \mathbf{x} \left(1 + \frac{1}{2}\mathbf{v}^2 - \frac{1}{2}U + \Pi\right) d^3x \tag{4.7.35}$$

moves with uniform velocity w.r.t. the global coordinate (harmonic or standard PN) system (Will 1981). This suggests defining the center of inertial mass of body a as:

$$\mathbf{x}_a = m_a^{-1} \int_a \rho^* \mathbf{x} \left(1 + \frac{1}{2}\bar{\mathbf{v}}^2 - \frac{1}{2}U_{(a)} + \Pi\right) d^3x \; . \tag{4.7.36}$$

Let us now further analyze the post-Newtonian Euler equations where we make use of the following relations, valid to PN order:

$$\int_a \rho(v^i U_{(a)} - V_{(a)}^i)\, d^3x = 0 \tag{4.7.37a}$$

$$\int_a \rho v^j V^j_{(a),i} \, d^3x = 0 \tag{4.7.37b}$$

$$\int_a \rho (\varphi U_{(a),i} + \Phi_{(a),i}) \, d^3x = 0 \tag{4.7.37c}$$

$$\int_a \rho v^j \partial_j (V^i_{(a)} - W^i_{(a)}) \, d^3x = 0 \ . \tag{4.7.37d}$$

Now, apart from terms linear in the PN potentials the total force F^i contains a non-linear part:

$$F^i_{\mathrm{NL}} = G \int_a \rho^* (\varphi U_{,i} + \Phi_{,i}) \, d^3x \ . \tag{4.7.38}$$

Whereas for the linear terms the splitting of the total force into self and external part is obvious, a corresponding splitting of F^i_{NL} is ambiguous; however, relation (4.7.37c) suggests defining the self-part of F^i_{NL} as

$$F^i_{\mathrm{NL(a)}} = G \int_a \rho^* (\varphi U_{(a),i} + \Phi_{(a),i}) \, d^3x \ , \tag{4.7.39}$$

since then the self-force vanishes to PN order. Note that $U_{(e)}$ is still contained in both terms of F^i_{NL} via φ. The global equation for the translational motion can therefore be written as

$$\frac{d}{dt} p^i = \frac{d}{dt} (p^i_{(a)} + p^i_{(e)}) = F^i_{(e)} \tag{4.7.40}$$

with

$$\mathbf{P}_{(a)} = \int_a \rho^* \left[\left(1 + \frac{1}{2} \mathbf{v}^2 - U_{(a)} + \Pi + p/\rho^* \right) \mathbf{v} + \frac{1}{2} (\mathbf{V}_{(a)} - \mathbf{W}_{(a)}) \right] d^3x$$

$$\mathbf{P}_{(e)} = \int_a \rho^* \left[(2\gamma + 1) \mathbf{v} U_{(e)} - (2\gamma + 2) \mathbf{V}_{(e)} + \frac{1}{2} (\mathbf{V}_{(e)} - \mathbf{W}_{(e)}) \right] d^3x$$

$$\tag{4.7.41}$$

and

$$F^i_{(e)} = G \int_a \left\{ \rho \partial_i U_{(e)} + \rho^* \left[-(2\gamma + 2) v^j V^j_{(e),i} + (\varphi U_{(e),i} + \Phi_{(e),i}) \right. \right.$$
$$\left. \left. + \frac{1}{2} v^j \partial_j (V^i_{(e)} - W^i_{(e)}) \right] \right\} d^3x \ . \tag{4.7.42}$$

It is tedious but straightforward to express the total momentum and total external force in terms of collective variables if i) simplifying assumptions about the fluid motions within the bodies are made and ii) higher multipole moments (quadrupole etc.) of the bodies are neglected.

Splitting the velocity field $\mathbf{v}$ inside body a according to

$$\mathbf{v}(t,\mathbf{x}) = \dot{\mathbf{x}}_a(t) + \bar{\mathbf{v}}(t,\mathbf{x}) \; , \tag{4.7.43}$$

the self-momentum can be written as

$$p_{(a)}^i = \left(1 + \frac{1}{2}\mathbf{v}_a^2\right) m_a v_a^i + \left(1 + \frac{1}{2}\mathbf{v}_a^2\right) D_{(a)}^i + v_a^i(\mathbf{v}_a \cdot \mathbf{D}_{(a)}) + \frac{1}{2}v_a^j \ddot{I}_{(a)}^{ij} \; , \tag{4.7.44}$$

where

$$\mathbf{D}_{(a)} = \int_a \rho^* \left(1 + \frac{1}{2}\bar{\mathbf{v}}^2 - \frac{1}{2}U_{(a)} + \Pi + p/\rho^*\right)\bar{\mathbf{v}} - \frac{1}{2}\mathbf{W}_{(a)}\right) d^3x$$
$$+ \frac{1}{2}\int_a \rho^* \left(\mathbf{V}_{(a)} - U_{(a)}\bar{\mathbf{v}}\right) d^3x \tag{4.7.45}$$

is a generalization of the Newtonian "current" dipole moment and $I_{(a)}^{ij}$ is the Newtonian inertia tensor:

$$I_{(a)}^{ij} = \int_a \rho^* r_{(a)}^i r_{(a)}^j \, d^3x \; . \tag{4.7.46}$$

Here we have made use of the Newtonian tensor virial relation that for each massive body reads

$$\frac{1}{2}\ddot{I}^{ij} = \int \rho^* \bar{v}^i \bar{v}^j - \frac{1}{2}\int \frac{\rho^* \rho^{*\prime}}{|\mathbf{x} - \mathbf{x}'|} \, d^3x \, d^3x' + \delta_{ij}\int p\, d^3x \; . \tag{4.7.47}$$

Now, the second term on the right hand side of (4.7.45) vanishes by virtue of the definition for $\mathbf{V}$, whereas the first term vanishes according to our post-Newtonian center of mass condition (4.7.36): differentiating (4.7.36) w.r.t. the time coordinate, using the continuity equation for ρ^* and the Newtonian equations of motion in any PN terms one finds (Will 1981):

$$m_a \mathbf{v}_a = \int_a \left[\rho^*\left(1 + \frac{1}{2}\bar{\mathbf{v}}^2 - \frac{1}{2}U_{(a)} + \Pi\right)\mathbf{v} + p\bar{\mathbf{v}} - \frac{1}{2}\rho^*\mathbf{W}\right] d^3x \tag{4.7.48}$$

which finally leads to the vanishing of $\mathbf{D}_a$.

If we furthermore assume the bodies to be *secularly stationary* in the sense that the (Newtonian) Virial theorem holds

$$\langle \ddot{I}^{ij} \rangle = 0 \; , \tag{4.7.49}$$

where the bracket denotes the time average over times large compared to internal dynamical time-scales, the self-momentum of a simply reduces to:

$$\mathbf{P}_{(a)} = \left(1 + \frac{1}{2}\mathbf{v}_a^2\right) m_a \mathbf{v}_a \; . \tag{4.7.50}$$

Similarly the external part of **p** and the external force can be evaluated. Neglecting higher multipole moments and terms of order r_{ab}^{-4} the final equation of motion can be derived from a Lagrangian with $(G = c = 1)$:

$$\mathcal{L} = \sum_a m_a \left(\frac{1}{2} \mathbf{v}_a^2 + \frac{1}{8} \mathbf{v}_a^4 \right)$$

$$+ \frac{1}{2} \sum_{\substack{a,b \\ b \neq a}} \frac{m_a^{(G)} m_b^{(G)}}{r_{ab}} \left[1 + (2\gamma + 1)\mathbf{v}_a^2 \right.$$

$$\left. - \frac{1}{2}(4\gamma + 3)\mathbf{v}_a \cdot \mathbf{v}_b - \frac{1}{2}(\mathbf{v}_a \cdot \mathbf{n}_{ab})(\mathbf{v}_b \cdot \mathbf{n}_{ab}) \right]$$

$$- \frac{(2\beta - 1)}{2} \sum_{\substack{a,b \\ b \neq a}} \sum_{c \neq a} \frac{m_a m_b}{r_{ab}} \frac{m_c}{r_{ac}}$$

$$+ \sum_{b \neq a} \left[\frac{(2\gamma + 1)}{2} m_b \frac{\mathbf{x}_{ab} \times \mathbf{v}_a}{r_{ab}^3} - (\gamma + 1)m_b \frac{\mathbf{x}_{ab} \times \mathbf{v}_b}{r_{ab}^3} \right] \mathbf{S}_a$$

$$- \sum_{b \neq a} \left[\frac{(2\gamma + 1)}{2} m_a \frac{\mathbf{x}_{ab} \times \mathbf{v}_b}{r_{ab}^3} - (\gamma + 1)m_a \frac{\mathbf{x}_{ab} \times \mathbf{v}_a}{r_{ab}^3} \right] \mathbf{S}_b$$

$$- \sum_{b \neq a} \left(\frac{\gamma + 1}{2} \right) \left[\frac{3(\mathbf{S}_a \cdot \mathbf{x}_{ab})(\mathbf{S}_b \cdot \mathbf{x}_{ab})}{r_{ab}^5} - \frac{\mathbf{S}_a \cdot \mathbf{S}_b}{r_{ab}^3} \right] \quad . \tag{4.7.51}$$

Here $\mathbf{S}_a$ denotes the spin (intrinsic angular momentum vector) of a, given to sufficient accuracy by:

$$\mathbf{S}_a = \int_a \rho \, \mathbf{r}_a \times \bar{\mathbf{v}} \, d^3 x \quad , \tag{4.7.52}$$

and $\mathbf{x}_{ab} = \mathbf{x}_b - \mathbf{x}_a$, $\hat{\mathbf{n}}_{ab} = \mathbf{x}_{ab}/r_{ab}$. The *gravitational mass* $m_a^{(G)}$ of a is given by

$$m_a^{(G)} = m_a \left(1 + \eta_N \frac{\Omega_a}{m_a} \right) \quad , \tag{4.7.53}$$

where the "Nordtvedt parameter" η_N reads

$$\eta_N = (4\beta - \gamma - 3) \tag{4.7.54}$$

and Ω_a denotes the *internal gravitational energy* of a:

$$\Omega_a = -\frac{1}{2} \int_a \rho^* U_{(a)} \, d^3 x \quad . \tag{4.7.55}$$

We first notice that if we neglect all internal degrees of freedom the Lagrangian (4.7.51) reduces to the corresponding Lagrangian (4.1.13) of the "point-mass" limit, justifying the derivation of the Einstein-Infeld-Hoffmann (EIH) equations of motion (4.1.14) by employing the PPN metric for a system of "point-masses".

The appearance of the η_N term in the gravitational mass is related to a violation of the "strong equivalence principle" (e.g. Will 1981) which extends the Einstein equivalence principle (Section 2.2) to the case of self-gravitating bodies.

Strong Equivalence Principle. *The weak equivalence principle is valid for self-gravitating bodies as well as for test bodies; the outcome of any local test experiment is independent of the velocity of the (freely falling) apparatus w.r.t. the fixed stars and where and when it is performed.*

For laboratory sized bodies the ratio of gravitational self-energy to rest energy is typically of order 10^{-26}, hence a test of the strong equivalence principle requires a study of the dynamics of macroscopic, astrophysical bodies. Equation (4.7.53) implies that in the case $\eta_N \neq 0$ the ratio of (passive) gravitational mass to inertial mass is different from unity:

$$m_a^{(G)}/m_a = 1 + \eta_N \Omega_a/m_a \ . \tag{4.7.56}$$

For the Earth-Moon system this leads to the important consequence that Earth and Moon would fall at different rates[†] towards the Sun causing an anomalous polarization of the Earth-Moon orbit (the "Nordtvedt effect"; Nordtvedt 1968, 1970, 1971, Will 1981). Writing Newton's law of gravity as

$$m_a \mathbf{a}_a = m_a^{(G)} \nabla U \tag{4.7.57}$$

(U containing the (active) gravitational mass), the accelerations of Moon and Earth in the field of the Sun are approximately given by

$$\mathbf{a}_\oplus = -\left(\frac{m^{(G)}}{m}\right)_\oplus \left[\frac{M_\oplus \mathbf{X}}{R^3} - \frac{m_\mathbb{C}^{(G)}\mathbf{x}}{r^3}\right]$$

$$\mathbf{a}_\mathbb{C} = -\left(\frac{m^{(G)}}{m}\right)_\mathbb{C} \left[\frac{M_\oplus(\mathbf{X}+\mathbf{x})}{|\mathbf{X}+\mathbf{x}|^3} + \frac{m_\oplus^{(G)}\mathbf{x}}{r^3}\right] \ ,$$

leading to

$$\mathbf{a} = \mathbf{a}_\mathbb{C} - \mathbf{a}_\oplus = -\frac{Gm^*\mathbf{x}}{r^3} + \eta_N\left[\left(\frac{\Omega}{m}\right)_\oplus - \left(\frac{\Omega}{m}\right)_\mathbb{C}\right]\frac{GM_\odot}{R^3}\mathbf{X}$$

$$+ \left(\frac{m^{(G)}}{m}\right)_\mathbb{C} GM_\odot\left(\frac{\mathbf{X}}{R^3} - \frac{\mathbf{X}+\mathbf{x}}{|\mathbf{X}+\mathbf{x}|^3}\right) \ , \tag{4.7.58}$$

---

† Since Ω is negative the Moon would fall faster towards the Sun than the Earth if $\eta_N > 0$; $-(\Omega/m)_\oplus \sim 5 \times 10^{-10}$, $-(\Omega/m)_\mathbb{C} \sim 2 \times 10^{-11}$.

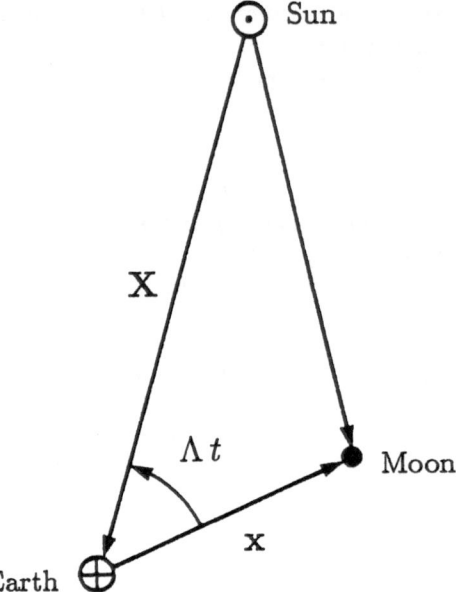

Fig. 4.5. Geometry in the problem of the Nordtvedt effect

where the last term indicates the usual Newtonian tidal perturbation of the lunar orbit caused by the Sun and

$$m^* = m_\oplus^{(G)} + m_{\text{\Moon}}^{(G)} + \eta_N \left[m_\oplus^{(G)} \left(\frac{\Omega}{m} \right)_{\text{\Moon}} + m_{\text{\Moon}}^{(G)} \left(\frac{\Omega}{m} \right)_\oplus \right] \qquad (4.7.59)$$

Here, $\mathbf{X}$ denotes the heliocentric vector of the geocenter and $\mathbf{x}$ the geocentric vector of the Moon's center of mass (Fig. 4.5). Assuming

$$\delta_N = \left(\frac{m^{(G)}}{m} \right)_{\text{\Moon}} - \left(\frac{m^{(G)}}{m} \right)_\oplus = \eta_N \left[\left(\frac{\Omega}{m} \right)_{\text{\Moon}} - \left(\frac{\Omega}{m} \right)_\oplus \right] \neq 0 \ , \qquad (4.7.60)$$

the additional acceleration of the Moon can be written as

$$\delta\mathbf{a} = -\frac{GM_\odot \mathbf{X}}{R^3} \delta_N \ . \qquad (4.7.61)$$

Assuming the Moon's unperturbed orbit about the Earth to be circular with angular velocity n and the Earth's orbit about the Sun as circular with frequency n' in the same plane, the radial equation for the lunar orbit reads:

$$\ddot{r} - \frac{h^2}{r^3} = -\frac{Gm^*}{r^2} + \delta a \cos \Lambda t \ , \qquad (4.7.62)$$

where **h** is the specific angular momentum of the Earth-Moon orbit

$$\mathbf{h} = \mathbf{x} \times \left(\frac{d\mathbf{x}}{dt}\right)$$

and $\Lambda = n - n'$. Since $d\mathbf{h}/dt = \mathbf{x} \times \delta\mathbf{a}$ the additional acceleration leads to

$$\dot{h} = -r\,\delta a\,\sin\Lambda t \;. \tag{4.7.63}$$

Linearizing about a circular orbit with $r = r_0 + \delta r$, $h = h_0 + \delta h$ (Will 1981) one finds:

$$\delta h = (r_0/\Lambda)\,\delta a\cos\Lambda t$$

and

$$\delta\ddot{r} + n^2\delta r = (1 + 2n/\Lambda)\,\delta a\cos\Lambda t$$

leading to

$$\delta r = \left(\frac{1 + 2n/\Lambda}{n^2 - \Lambda^2}\right)\delta a\cos\Lambda t$$

$$\equiv \eta_N A_N\cos\Lambda t \simeq 8.0\,\eta_N\cos\Lambda t\;[\mathrm{m}]\;. \tag{4.7.64}$$

Including tidal effects the value for the Nordtvedt amplitude A_N is modified to $\sim 9.2\,\mathrm{m}$ (for more details see Will 1981). Lunar laser ranging gave the following limits for the Nordtvedt parameter η_N:

$$\eta_N = \begin{cases} 0.00 \pm 0.03 & \text{(Williams et al. (1976))} \\ 0.001 \pm 0.015 & \text{(Shapiro et al. (1976))} \end{cases},$$

where realistic 1σ-errors are given.

Let us finally come to the spin-dependent terms in the Lagrangian (4.7.51). In the context of post-Newtonian formalism they have been derived by Brumberg (1972) who also discusses the physical implications in detail (see also Kalitzin 1959, Michalska 1960). They have also been derived using a one-graviton exchange theory by Barker et al. (1975) and applied to the motion of a binary system (Barker et al. 1975, 1976, 1981, 1982, 1986; see also Damour et al. 1988). The two lines before the last one in (4.7.51) represent the spin-orbit coupling and the last line the spin-spin coupling in analogy to the situation in a quantum mechanical system. For a binary system the spin-orbit interaction potential V_{S-O} can be written as ($\mathbf{x} = \mathbf{x}_1 - \mathbf{x}_2$):

$$V_{S-O} = V_{S_1} + V_{S_2} \tag{4.7.65}$$

with

$$V_{S_1} = \frac{G}{c^2 r^3}\left[(\gamma + 1) + \left(\gamma + \frac{1}{2}\right)\frac{m_2}{m_1}\right]\mathbf{S}_1 \cdot (\mathbf{x} \times \mathbf{p}) \tag{4.7.66a}$$

$$V_{S_2} = \frac{G}{c^2 r^3}\left[(\gamma + 1) + \left(\gamma + \frac{1}{2}\right)\frac{m_1}{m_2}\right]\mathbf{S}_2 \cdot (\mathbf{x} \times \mathbf{p})\;, \tag{4.7.66b}$$

with $\mathbf{p} = m_1\mathbf{v}_1 = -m_2\mathbf{v}_2$ to sufficient accuracy, and the spin-spin interaction potential reads:

$$V_{S-S} = \left(\frac{\gamma + 1}{2}\right) \frac{G}{c^2 r^3} \left[\frac{3(\mathbf{S}_1 \cdot \mathbf{x})(\mathbf{S}_2 \cdot \mathbf{x})}{r^2} - \mathbf{S}_1 \cdot \mathbf{S}_2\right] . \tag{4.7.67}$$

The perturbing acceleration due to the spin-orbit coupling for the relative motion in a binary system reads:

$$\mathbf{a}_{S-O} = \mathbf{v} \times (\nabla \times \boldsymbol{\zeta}) \tag{4.7.68}$$

with

$$\boldsymbol{\zeta} = 2G \frac{\mathbf{x} \times \mathbf{J}^*}{r^3} \tag{4.7.69}$$

where

$$\mathbf{J}^* = \frac{1}{2}\left(\frac{1 + 2\gamma}{2}\frac{m_2}{m_1} + (\gamma + 1)\right)\mathbf{S}_1 + \frac{1}{2}\left(\frac{1 + 2\gamma}{2}\frac{m_1}{m_2} + (\gamma + 1)\right)\mathbf{S}_2 . \tag{4.7.70}$$

This generalizes the Lense-Thirring acceleration $\mathbf{a}_{LT}$ (4.2.16) to the case of two rotating bodies. For $\mathbf{S}_1 = 0$ and $m_1/m_2 \ll 1$ the spin-orbit acceleration agrees with $\mathbf{a}_{LT}$. In the general case a comparison of (4.2.16) with (4.7.68) shows that all results from Section 4.2.2 apply provided we replace $(\gamma + 1)\mathbf{J}/2$ by $\mathbf{J}^*$. For example for the secular perturbations we find to first order

$$\left\langle\frac{d\Omega}{dt}\right\rangle = \frac{\xi^*}{(1 - e^2)^{3/2}} \tag{4.7.71a}$$

$$\left\langle\frac{d\tilde{\omega}}{dt}\right\rangle = \left\langle\frac{de}{dt}\right\rangle = (1 - 3\cos I)\left\langle\frac{d\Omega}{dt}\right\rangle \tag{4.7.71b}$$

with

$$\xi^* = \frac{2GJ^*}{na^3(1 - e^2)^{3/2}} .$$

The perturbing acceleration for the relative motion in a binary system due to the spin-spin coupling is immediately obtained from $\mu^{-1}\partial\mathcal{L}_{S-S}/\partial\mathbf{x}$ and is given by:

$$\mathbf{a}_{S-S} = -\frac{3GS_1S_2}{a^5\mu}\left(\frac{\gamma + 1}{2}\right)\frac{a^5}{r^4}[(\mathbf{n}_2 \cdot \mathbf{n})\mathbf{n}_1 + (\mathbf{n}_1 \cdot \mathbf{n})\mathbf{n}_2$$
$$- 5(\mathbf{n}_1 \cdot \mathbf{n})(\mathbf{n}_2 \cdot \mathbf{n})\mathbf{n} + (\mathbf{n}_1 \cdot \mathbf{n}_2)\mathbf{n}] , \tag{4.7.72}$$

where $\mathbf{n}_i$ are Euclidean unit vectors in the direction of $\mathbf{S}_i$. The decomposition of $\mathbf{a}_{S-S}$ follows immediately from (4.2.5) which can be used to derive the general perturbations of the orbital elements. The secular perturbations, which are of major interest, can also be found in the following way (Brumberg 1972) by

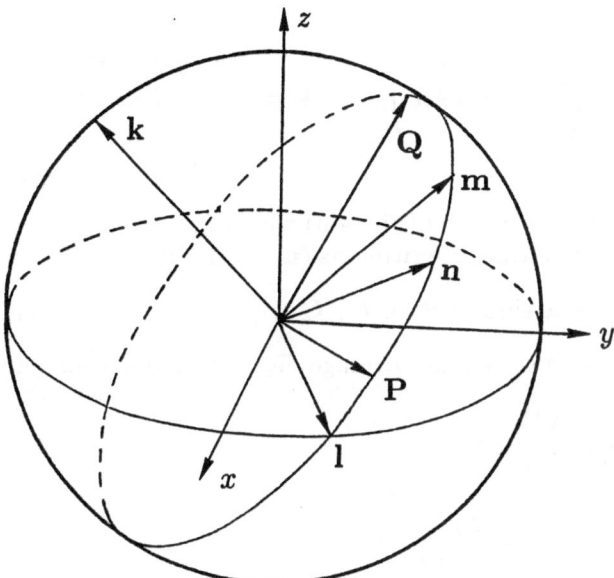

Fig. 4.6. Some useful Euclidean systems of triad vectors

employing the perturbation equations in the usual form with the perturbing function R. For the relative binary motion R is given by

$$L = \mathcal{L}/\mu = \frac{1}{2}\mathbf{v}^2 + \frac{Gm}{r} + R \tag{4.7.73}$$

and the spin-spin contribution reads:

$$
\begin{aligned}
R_{S-S} = {} & \frac{GS_1 S_2}{a^3 \mu} \left(\frac{\gamma+1}{2}\right) \left(\frac{a}{r}\right)^3 [(\mathbf{n}_1 \cdot \mathbf{n}_2) - 3(\mathbf{n}_1 \cdot \mathbf{n})(\mathbf{n}_2 \cdot \mathbf{n})] \\
= {} & \frac{GS_1 S_2}{a^3 \mu} \left(\frac{\gamma+1}{2}\right) \Big[[\mathbf{n}_1 \cdot \mathbf{n}_2 \\
& - \frac{3}{2}(\mathbf{n}_1 \cdot \mathbf{P})(\mathbf{n}_2 \cdot \mathbf{P}) - \frac{3}{2}(\mathbf{n}_1 \cdot \mathbf{Q})(\mathbf{n}_2 \cdot \mathbf{Q})\Big] \left(\frac{a}{r}\right)^3 \\
& + \frac{3}{2}[(\mathbf{n}_1 \cdot \mathbf{Q})(\mathbf{n}_2 \cdot \mathbf{Q}) - (\mathbf{n}_1 \cdot \mathbf{P})(\mathbf{n}_2 \cdot \mathbf{P})] \cos 2f \left(\frac{a}{r}\right)^3 \\
& - \frac{3}{2}[(\mathbf{n}_1 \cdot \mathbf{P})(\mathbf{n}_2 \cdot \mathbf{Q}) - (\mathbf{n}_1 \cdot \mathbf{Q})(\mathbf{n}_2 \cdot \mathbf{P})] \sin 2f \left(\frac{a}{r}\right)^3 \Big] \ , \tag{4.7.74}
\end{aligned}
$$

where $\mathbf{P}$ and $\mathbf{Q}$ are Euclidean unit vectors in the orbital plane, with $\mathbf{P}$ pointing towards perihelion:

$$\mathbf{P} = \mathbf{l}\cos\omega + \mathbf{m}\sin\omega \ ; \quad \mathbf{Q} = -\mathbf{l}\sin\omega + \mathbf{m}\cos\omega \ . \tag{4.7.75}$$

Here

$$
\mathbf{l} = \begin{pmatrix} \cos\Omega \\ \sin\Omega \\ 0 \end{pmatrix} \quad ; \quad \mathbf{m} = \begin{pmatrix} -\cos I \sin\Omega \\ +\cos I \cos\Omega \\ \sin I \end{pmatrix} \quad ; \quad \mathbf{k} = \begin{pmatrix} +\sin I \sin\Omega \\ -\sin I \cos\Omega \\ \cos I \end{pmatrix}
$$

(4.7.76)

constitute a Euclidean triad of unit vectors (Fig. 4.6) with $\mathbf{l}$ directed along the line of nodes. Note, that $\mathbf{n} = \mathbf{x}/r$ can be written as ($u = \omega + f$):

$$
\mathbf{n} = \mathbf{l}\cos u + \mathbf{m}\sin u = \mathbf{P}\cos f + \mathbf{Q}\sin f \ . \tag{4.7.77}
$$

To derive the secular perturbations we can average R_{S-s} over one complete revolution with

$$
\left\langle \left(\frac{a}{r}\right)^3 \right\rangle = (1 - e^2)^{-3/2}
$$

and

$$
\left\langle \left(\frac{a}{r}\right)^3 \cos 2f \right\rangle = \left\langle \left(\frac{a}{r}\right)^3 \sin 2f \right\rangle = 0 \ .
$$

One finds that

$$
\langle R_{S-s} \rangle = \frac{G S_1 S_2}{a^3 (1 - e^2)^{3/2} \mu} \left(\frac{\gamma + 1}{2}\right) \Big[\mathbf{n}_1 \cdot \mathbf{n}_2
$$
$$
- \frac{3}{2}(\mathbf{n}_1 \cdot \mathbf{l})(\mathbf{n}_2 \cdot \mathbf{l}) - \frac{3}{2}(\mathbf{n}_1 \cdot \mathbf{m})(\mathbf{n}_2 \cdot \mathbf{m}) \Big] \tag{4.7.78}
$$

and the secular perturbations can be obtained directly from $\langle R_{S-s} \rangle$ via the perturbation equations by differentiation w.r.t. the orbital elements, e.g.

$$
\left\langle \frac{d\Omega}{dt} \right\rangle = \frac{1}{na^2 \sqrt{1 - e^2} \sin I} \frac{\partial \langle R_{S-s} \rangle}{\partial I}
$$
$$
= -\frac{3}{2} \frac{G S_1 S_2}{na^5 (1 - e^2)^2 \mu \sin I} [(\mathbf{n}_1 \cdot \mathbf{k})(\mathbf{n}_2 \cdot \mathbf{m})
$$
$$
+ (\mathbf{n}_1 \cdot \mathbf{m})(\mathbf{n}_2 \cdot \mathbf{k})] \tag{4.7.79a}
$$

since $\partial \mathbf{p} / \partial I = \mathbf{k}$. Similarly one finds (Brumberg 1972):

$$
\left\langle \frac{dI}{dt} \right\rangle = -\frac{3}{2} \frac{G S_1 S_2}{na^5 (1 - e^2)^2 \mu} [(\mathbf{n}_1 \cdot \mathbf{k})(\mathbf{n}_2 \cdot \mathbf{l})
$$
$$
+ (\mathbf{n}_1 \cdot \mathbf{l})(\mathbf{n}_2 \cdot \mathbf{k})] \tag{4.7.79b}
$$

$$
\left\langle \frac{d\omega}{dt} \right\rangle = +3 \frac{G S_1 S_2}{na^5 (1 - e^2)^2 \mu \sin I} \Big[(\mathbf{n}_1 \cdot \mathbf{n}_2)
$$
$$
- \frac{3}{2}(\mathbf{n}_1 \cdot \mathbf{l})(\mathbf{n}_2 \cdot \mathbf{l}) - \frac{3}{2}(\mathbf{n}_1 \cdot \mathbf{m})(\mathbf{n}_2 \cdot \mathbf{m}) \Big] \tag{4.7.79c}
$$

$$
\left\langle \frac{d\mathcal{M}}{dt} \right\rangle = (1 - e^2)^{1/2} \left\langle \frac{d\omega}{dt} \right\rangle \tag{4.7.79d}
$$

where
$$\frac{dM}{dt} = n + \frac{d\mathcal{M}}{dt} \tag{4.7.80}$$

and
$$\left\langle \frac{da}{dt} \right\rangle = \left\langle \frac{de}{dt} \right\rangle = 0 \ . \tag{4.7.81}$$

4.7.2 Spin Motion

There is a variety of ways to analyze the motion of the spins e.g. for a binary system. One way is to decompose the total angular momentum vector of the system

$$L_i = \epsilon_{ijk} \int x^j \pi^k \, d^3x = \epsilon_{ijk} \int x^j \Theta^{k0} \, d^3x \ , \tag{4.7.82}$$

where π^k is again given by (4.7.15) and $\Theta^{\mu\nu}$ is the stress energy complex (4.7.23), into the orbital angular momentum $\mathbf{J}$ (A2.9) from the "point-mass" limit and the spins of the individual bodies. Since $\mathbf{L}$ is a conserved quantity the time derivative of $\mathbf{J}$ is related with those of the spins. The *spin motion* in a binary system can therefore be found from an analysis of the *orbital motion*. For quasi-rigid motion one finds that the motion of the spins can directly be inferred from the Lagrangian (4.7.51) considered as a function of the angular velocities of the two bodies (Barker et al. 1970). For body 1 one finds (similar results apply for body 2):

$$\frac{d}{dt} \mathbf{n}_1 = \mathbf{\Omega}^{(1)} \times \mathbf{n}_1 \tag{4.7.83}$$

with
$$\mathbf{\Omega}^{(1)} = \mathbf{\Omega}^{(1)}_{S-O} + \mathbf{\Omega}^{(1)}_{S-S} \tag{4.7.84}$$

where the spin-orbit (de Sitter-Fokker) part reads:

$$\mathbf{\Omega}^{(1)}_{S-O} = \frac{G\mu}{c^2 r^3} \left[(\gamma + 1) + \left(\gamma + \frac{1}{2} \right) \frac{m_2}{m_1} \right] \mathbf{x} \times \mathbf{v} \tag{4.7.85}$$

and the spin-spin (generalized Lense-Thirring or Schiff) contribution is given by:

$$\mathbf{\Omega}^{(1)}_{S-S} = \frac{G}{c^2 r^3} \left(\frac{\gamma + 1}{2} \right) \left(\frac{3\mathbf{x}(\mathbf{S}_2 \cdot \mathbf{x})}{r^2} - \mathbf{S}_2 \right) \ . \tag{4.7.86}$$

Averaging the precession frequency $\mathbf{\Omega}^{(1)}$ over one complete revolution gives the secular drift of $\mathbf{n}_1$ (relativistic spin precession) as

$$\left\langle \frac{d}{dt} \mathbf{n}_1 \right\rangle = \left\langle \mathbf{\Omega}^{(1)} \right\rangle \times \mathbf{n}_1 \tag{4.7.87}$$

with
$$\left\langle \mathbf{\Omega}^{(1)}_{S-O} \right\rangle = \frac{G\mu n}{c^2 a(1 - e^2)} \left[(\gamma + 1) + \left(\frac{2\gamma + 1}{2} \right) \frac{m_2}{m_1} \right] \mathbf{k} \tag{4.7.88}$$

and

$$\left\langle \mathbf{\Omega}_{S-s}^{(1)} \right\rangle = \frac{GS_2}{2c^2 a^3 (1-e^2)^{3/2}} \left(\frac{\gamma+1}{2} \right) [\mathbf{n}_2 - 3(\mathbf{n}_2 \cdot \mathbf{k})\mathbf{k}] \tag{4.7.89}$$

with $n^2 a^3 = Gm$.

4.7.3 Beyond the First Post-Newtonian Approximation

The problems of gravitational radiation and the dynamics of the binary pulsar have led to the necessity of describing systems of gravitationally interacting bodies (especially close binary systems) at higher than the first post-Newtonian order; up to the second PN level the total energy of an N-body system is conserved: there is no emission of gravitational radiation. Several formulations of higher post-Newtonian approximation (PNA) schemes exist in the literature that differ i) by the way the expansion is carried out and ii) by the gauge condition used. The foundations of present PNA theories were laid by Droste (1916), de Sitter (1916), Lorentz et al. (1917), Chazy (1928, 1930), Levi-Civita (1937, 1950), Einstein et al. (1938), Eddington et al. 1938, Papapetrou (1951), Infeld et al. (1960), Fock (1964), Chandrasekhar (1965) (PNA1), Chandrasekhar et al. (1969) (PNA2), Chandrasekhar et al. (1970) (PNA2.5), Ohta et al. (1973, 1974), Spyrou (1978). A thorough and detailed review of the subject has recently been presented by Damour (1987a) upon which the following is partially based.

To have at least formally a concept of a post-Newtonian hierarchy one can start with Einstein's field equations in the Landau-Lifshitz form (Landau et al. 1962, Misner et al. 1973):

$$[\mathbf{g}^{\mu\nu}\mathbf{g}^{\rho\sigma} - \mathbf{g}^{\mu\rho}\mathbf{g}^{\nu\sigma}]_{,\rho\sigma} = 16\pi G\Theta^{\mu\nu} \ , \tag{4.7.90}$$

where $\mathbf{g}^{\mu\nu}$ is given in (4.7.2) ($\mathbf{g}^{\mu\nu} = \sqrt{-g}\,g^{\mu\nu}$) and $\Theta^{\mu\nu}$ is the stress energy complex from (4.7.23). After imposing e.g. the harmonic gauge condition (4.7.2) this leads to the "reduced field equations" in the form

$$\Box_\eta \gamma^{\mu\nu} = 16\pi G\Lambda^{\mu\nu} \ , \tag{4.7.91}$$

where $\Box_\eta$ denotes the flat space d'Alembertian

$$\Box_\eta = \eta^{\rho\sigma}\partial_\rho\partial_\sigma \tag{4.7.92}$$

$$\Lambda^{\mu\nu} = \Theta^{\mu\nu} - \frac{1}{16\pi G}[\gamma^{\mu\nu}\gamma^{\rho\sigma} - \gamma^{\mu\rho}\gamma^{\nu\sigma}]_{,\rho\sigma} \tag{4.7.93}$$

and we have written

$$\mathbf{g}^{\mu\nu} = \eta^{\mu\nu} + \gamma^{\mu\nu} \tag{4.7.94}$$

in the harmonic coordinate system $(t, \mathbf{x})$. Now, generally one assumes $\gamma^{\mu\nu}$ to be small everywhere and to admit an asymptotic expansion, say, in terms of

ξ, with $\xi^2 = GM/c^2 L$ being the maximum value of $Gm/c^2 r$ in the (isolated) system of the type (weak field approximation):

$$\gamma^{\mu\nu}(t, \mathbf{x}) = \xi^2 \, \gamma^{\mu\nu}_{(2)}(t, \mathbf{x}) + \xi^3 \, \gamma^{\mu\nu}_{(3)}(t, \mathbf{x}) + \cdots + \xi^n \, \gamma^{\mu\nu}_{(n)}(t, \mathbf{x}) + \cdots . \quad (4.7.95)$$

One furthermore assumes that (slow motion approximation)

$$\left| \frac{1}{c} \frac{\partial \gamma^{\mu\nu}}{\partial t} \right| \sim \xi \left| \frac{\partial \gamma^{\mu\nu}}{\partial x^i} \right| . \quad (4.7.96)$$

As we have already seen in Chapter 2 the (reduced) field equations determine the various $\gamma^{\mu\nu}_{(n)}$ uniquely only after a suitable choice of boundary conditions, such as (2.1.23):

$$\lim_{\substack{|\mathbf{x}| \to \infty \\ t = \text{const.}}} \gamma^{\mu\nu}_{(n)} = 0 . \quad (4.7.97)$$

This program then leads to a formal hierarchy of Poisson equations for the $\gamma^{\mu\nu}_{(n)}$s of the form:

$$\Delta(\xi^n \gamma^{\mu\nu}_{(n)}) = T^{\mu\nu} - \text{terms} + (\text{terms known from preceeding approximations}) .$$

However, terms resulting from lower approximations lead to badly behaved Poisson equations which do not admit any solutions fulfilling the boundary conditions (4.7.97). For the harmonic gauge this happens already at the PNA2-level (i.e. for γ^{00} to $\mathcal{O}(\epsilon^6)$ and γ^{0i} to $\mathcal{O}(\epsilon^5)$). The problem lies not so much in the gauge condition but in the choice of the boundary condition (4.7.97): the slow motion weak field approximation (4.7.95, 4.7.96) is basically a *near-zone* expansion of the exact $\gamma^{\mu\nu}(t, \mathbf{x})$, valid for $r \ll \lambda$, where λ is a characteristic wavelength of the gravitational radiation emitted by the system. Having this in mind one can chose a different boundary condition for the hierarchy by transforming first (4.7.91) for $\gamma^{\mu\nu}$ into an integral equation by means of the retarded Green's function assuming the absence of incoming gravitational radiation :

$$\gamma^{\mu\nu}(t, \mathbf{x}) = 4G \int \frac{\delta(x^0 - x^{0\prime} - |\mathbf{x} - \mathbf{x}'|)}{|\mathbf{x} - \mathbf{x}'|} \Lambda^{\mu\nu}(x') \, d^4 x'$$

$$= 4G \int \frac{\Lambda^{\mu\nu}(x^0 - |\mathbf{x} - \mathbf{x}'|)}{|\mathbf{x} - \mathbf{x}'|} \, d^3 x' . \quad (4.7.98)$$

Expanding $\Lambda^{\mu\nu}$ in a Taylor series about x^0 one obtains formally

$$\gamma^{\mu\nu}(t, \mathbf{x}) = 4G \sum_{n=0}^{\infty} \frac{(-1)^n}{n! \, c^n} \frac{\partial^n}{\partial t^n} \int \Lambda^{\mu\nu}(t, \mathbf{x}') |\mathbf{x} - \mathbf{x}'|^{n-1} \, d^3 x' . \quad (4.7.99)$$

The various $\gamma^{\mu\nu}_{(n)}$s can then be generated by successive approximation. This is the PNA hierarchy as suggested by Anderson et al. (1975) based on works by Peres (1959, 1960), Carmeli (1964, 1965), Synge (1969, 1970), Hogan et al. (1974) and others. It soon turned out that this PNA hierarchy is also not free of divergencies appearing at the 2.5 level, where radiation reaction terms first come into play in the equations of motion. Ehlers (1980) improved the Anderson-Decanio scheme essentially by leaving the time derivatives inside the integral in (4.7.99) and 'reducing them' by means of the equations of motion. As was shown by Kerlick (1980) this Ehlers scheme is free of divergencies up to and including the PNA2.5 level. However, the use of the near-zone expanded retarded potentials will cause the appearance of divergent integrals at least at the PNA4 level, corresponding to ξ^{10} or c^{-10} in g_{00} (Blanchet et al. 1988). The reason for this lies in the space-time non-locality of the gravitational interaction basically caused by the non-linearity ('gravity generates gravity') of the field equations and the fact the the gravitational field cannot propagate faster than light. The physical picture of this non-locality is that gravity waves emitted from the system can back-scatter off the curvature of space-time and influence the motion of the system at later times. This leads to retarded correlations over arbitrarily large time spans in the dynamics of gravitating systems and to the impossibility of expressing the metric tensor in terms of the *instantaneous* state of the matter variables. This happens first for the PNA4 level (Blanchet et al. 1988), where any post-Newtonian scheme eventually will break down.

Besides these PNA schemes two alternative approximation methods are currently available to obtain the desired equations of motion for massive bodies:

— the 'point particle approach' and
— the 'matched asymptotic expansion' technique.

In the first method the bodies are idealized to point particles with the consequence that regularization procedures have to be introduced to give the divergent functions on the world-line of particles a well defined meaning. In principle the results might depend upon the regularization scheme used. On the other hand the point particle technique seems to be very useful to derive the equations of motion of higher post-Newtonian order.

The method of asymptotic matching is based on matching the solutions valid near and far the sources. So far it is the *only* technique that is able to deal with the motion of gravitationally *condensed bodies* (neutron stars, black holes).

Explicit results for the motion of N point-masses at the second post-Newtonian level (PNA2) have been obtained by Ohta et al. (1974) (for corrections see Damour 1982, 1983, Damour et al. 1985b) in the ADM-gauge and by Damour et al. (1985) in harmonic gauge. Schäfer (1985) derived the full PNA2.5 dynamics of an N-body system in ADM gauge for point particles. The explicit Hamiltonian for the PNA2 three-body system is given in Schäfer (1987).

The full PNA2.5 two-body equations of motion valid even for condensed strongly self-gravitating bodies have first been derived in Damour et al. (1981)

and Damour (1982) (harmonic gauge) using the asymptotic matching technique. These results have been confirmed by Kopejkin (1985) and Grishchuk et al. (1986) using the Ehlers scheme discussed above for two weakly self-gravitating fluid balls. In harmonic coordinates the acceleration of the first body in the two-body system in the PNA2.5 approximation can be written as (Damour et al. 1981a,b):

$$\mathbf{a} = \mathbf{a}_0 + c^{-2}\mathbf{a}_2 + c^{-4}\mathbf{a}_4 + c^{-5}\mathbf{a}_5 + \mathcal{O}(\epsilon^6) \ , \tag{4.7.100}$$

where

$$\mathbf{a}_0 = -\frac{Gm_2\mathbf{n}}{r^2}$$

is the Newtonian and

$$\mathbf{a}_2 = \frac{Gm_2}{r^2}\left\{\mathbf{n}\left[-\mathbf{v}_1^2 - 2\mathbf{v}_2^2 + 4(\mathbf{v}_1\mathbf{v}_2) + \frac{3}{2}(\mathbf{nv}_2)^2 + 5\left(\frac{Gm_1}{r}\right) + 4\left(\frac{Gm_2}{r}\right)\right]\right.$$
$$\left. + (\mathbf{v}_1 - \mathbf{v}_2)\left[4(\mathbf{nv}_1) - 3(\mathbf{nv}_2)\right]\right\} \tag{4.7.101}$$

the post-Newtonian acceleration of the first body for $\beta = \gamma = 1$ (see A2). The PNA2 and PNA2.5 accelerations of body 1 are given by (Damour et al. 1981a, Damour 1987):

$$\mathbf{a}_4 = \frac{Gm_2}{r^2}\left\{\mathbf{n}\left[-2\mathbf{v}_2^4 + 4\mathbf{v}_2^2(\mathbf{v}_1\mathbf{v}_2) - 2(\mathbf{v}_1\mathbf{v}_2)^2 + \frac{3}{2}\mathbf{v}_1^2(\mathbf{nv}_2)^2\right.\right.$$
$$+ \frac{9}{2}\mathbf{v}_2^2(\mathbf{nv}_2)^2 - 6(\mathbf{v}_1\mathbf{v}_2)(\mathbf{nv}_2)^2 - \frac{15}{8}(\mathbf{nv}_2)^4$$
$$+ \left(\frac{Gm_1}{r}\right)\left(-\frac{15}{4}\mathbf{v}_1^2 + \frac{5}{4}\mathbf{v}_2^2 - \frac{5}{2}(\mathbf{v}_1\mathbf{v}_2)\right.$$
$$+ \frac{39}{2}(\mathbf{nv}_1)^2 - 39(\mathbf{nv}_1)(\mathbf{nv}_2) + \frac{17}{2}(\mathbf{nv}_2)^2\right)$$
$$\left. + \left(\frac{Gm_2}{r}\right)\left(4\mathbf{v}_2^2 - 8(\mathbf{v}_1\mathbf{v}_2) + 2(\mathbf{nv}_1)^2 - 4(\mathbf{nv}_1)(\mathbf{nv}_2) - 6(\mathbf{nv}_2)^2\right)\right]$$
$$+ (\mathbf{v}_1 - \mathbf{v}_2)[\mathbf{v}_1^2(\mathbf{nv}_2) + 4\mathbf{v}_2^2(\mathbf{nv}_1) - 5\mathbf{v}_2^2(\mathbf{nv}_2) - 4(\mathbf{v}_1\mathbf{v}_2)(\mathbf{nv}_1)$$
$$+ 4(\mathbf{v}_1\mathbf{v}_2)(\mathbf{nv}_2) - 6(\mathbf{nv}_1)(\mathbf{nv}_2)^2 + \frac{9}{2}(\mathbf{nv}_2)^3$$
$$\left. + \left(\frac{Gm_1}{r}\right)\left(-\frac{63}{4}(\mathbf{nv}_1) + \frac{55}{4}(\mathbf{nv}_2)\right) + \left(\frac{Gm_2}{r}\right)\left(-2(\mathbf{nv}_1) - 2(\mathbf{nv}_2)\right)\right]\right\}$$
$$+ \frac{G^3m_2}{r^4}\mathbf{n}\left[-\frac{57}{4}m_1^2 - 9m_2^2 - \frac{69}{2}m_1m_2\right] \tag{4.7.102}$$

and $(\mathbf{v} = \mathbf{v}_1 - \mathbf{v}_2)$

$$\mathbf{a}_5 = \frac{4}{5}\frac{G^2m_1m_2}{r^3}\left\{\mathbf{v}\left[-\mathbf{v}^2 + 2\left(\frac{Gm_1}{r}\right) - 8\left(\frac{Gm_2}{r}\right)\right]\right.$$
$$\left. + \mathbf{n}(\mathbf{nv})\left[3\mathbf{v}^2 - 6\left(\frac{Gm_1}{r}\right) + \frac{52}{3}\left(\frac{Gm_2}{r}\right)\right]\right\} \tag{4.7.103}$$

Using the *Newtonian* center of mass conditions the PNA2 and PNA2.5 *relative* accelerations $\mathbf{a}_4$ and $\mathbf{a}_5$ are given by:

$$
\begin{aligned}
\mathbf{a}_4 = {} & \frac{Gm\mathbf{n}}{r^2}\nu\left[-2v^4 + \frac{3}{2}v^2(\mathbf{n}\cdot\mathbf{v})^2(3-4\nu) - \frac{15}{8}(\mathbf{n}\cdot\mathbf{v})^4(1-3\nu)\right] \\
& + \frac{G^2m^2\mathbf{n}}{r^3}\left[\frac{1}{2}v^2\nu(11+4\nu) + 2(\mathbf{n}\cdot\mathbf{v})^2[1+\nu(12+3\nu)]\right] \\
& + \frac{Gm\mathbf{v}}{r^2}\nu\left[8v^2(\mathbf{n}\cdot\mathbf{v}) - \frac{3}{2}(\mathbf{n}\cdot\mathbf{v})^3(3+2\nu)\right] \\
& - \frac{1}{2}\frac{G^2m^2}{r^3}\mathbf{v}(\mathbf{n}\cdot\mathbf{v})(4+43\nu) - \frac{G^3m^3\mathbf{n}}{r^4}\left(9+\frac{87}{4}\nu\right)
\end{aligned}
\tag{4.7.104}
$$

and

$$
\mathbf{a}_5 = -\frac{8}{5}\frac{G^2m^2}{r^3}\nu\left\{\mathbf{v}\left[v^2 + 3\frac{Gm}{r}\right] - \mathbf{n}(\mathbf{n}\cdot\mathbf{v})\left[3v^2 + \frac{17}{3}\frac{Gm}{r}\right]\right\} . \tag{4.7.105}
$$

It is noteworthy that if we drop the $\mathbf{a}_5$ term the total acceleration can be obtained from a generalized Lagrangian that depends not only upon positions and velocities but also on *accelerations* (Damour et al. 1981b, Damour 1982) allowing the construction of the usual ten Noetherian constants of motion (energy, momentum, angular momentum, center of mass).

The usual S, T, W decomposition of $\mathbf{a}_5$ yields in terms of Keplerian (coordinate) osculating elements ($n_0^2 a^3 = GM$):

$$
\begin{aligned}
S = {} & +\frac{8}{5}\frac{G^3m^3n_0}{c^5a^3(1-e^2)^{9/2}}\nu(1+e\cos f)^3 e\sin f \\
& \times\left[\frac{14}{3} + 2e^2 + \frac{20}{3}e\cos f\right]
\end{aligned}
\tag{4.7.106a}
$$

$$
T = -\frac{8}{5}\frac{G^3m^3n_0}{c^5a^3(1-e^2)^{9/2}}\nu(1+e\cos f)^4(4+e^2+5e\cos f) \tag{4.7.106b}
$$

$$
W = 0 \tag{4.7.106c}
$$

leading to secular changes of the (Newtonian) semi-major axis and eccentricity of

$$
\left\langle\frac{da}{dt}\right\rangle = -\frac{64}{5}\frac{G^3m_1m_2(m_1+m_2)}{c^5a^3(1-e^2)^{7/2}}\left(1+\frac{73}{24}e^2+\frac{37}{96}e^4\right) \tag{4.7.107a}
$$

$$
\left\langle\frac{de}{dt}\right\rangle = -\frac{304}{15}\frac{G^3m_1m_2(m_1+m_2)e}{c^5a^4(1-e^2)^{5/2}}\left(1+\frac{121}{304}e^2\right) . \tag{4.7.107b}
$$

The secular decrease of the semi-major axis is obviously related to a loss of (Noetherian) energy of the two-body system due to the emission of gravitational waves. It leads to a secular change of the orbital period $\mathcal{P}$ according to

$$
\dot{\mathcal{P}} = \frac{3}{2}\left(\frac{a^3}{m}\right)^{1/2}\left(\frac{\dot{a}}{a}\right) = -\frac{3}{2}\mathcal{P}E^{-1}\left\langle\frac{dE}{dt}\right\rangle ,
$$

where E denotes the Noetherian energy of the PNA2 problem (Damour 1983a, 1987), of

$$\dot{\mathcal{P}} = -\frac{192\pi}{5c^5} \left(\frac{2\pi G}{\mathcal{P}}\right)^{5/3} \frac{m_1 m_2}{(m_1 + m_2)^{1/3}} F(e) \tag{4.7.108}$$

where

$$F(e) = \left(1 + \frac{73}{24}e^2 + \frac{37}{96}e^4\right)(1 - e^2)^{-7/2} . \tag{4.7.109}$$

Now, the secular change of the orbital period can be related with the power of gravitational radiation emitted to infinity. One has good reason to believe that this power emitted in form of gravitational radiation is given by the *gravitational flux formula* (e.g. Landau et al. 1962, Misner et al. 1973):

$$\frac{dE_{\mathrm{rad}}}{dt} = -\frac{G}{5c^4} \langle \dddot{\mathcal{I}}_{ij} \dddot{\mathcal{I}}_{ij} \rangle , \tag{4.7.110}$$

where $\mathcal{I}_{ij}$ is the trace-free moment of inertia tensor, to lowest order given by

$$\mathcal{I}_{ij} = \int \rho(t, \mathbf{x}) \left(x_i x_j - \frac{1}{3}\delta_{ij} r^2\right) d^3 x . \tag{4.7.111}$$

Analyzing the quadrupole flux formula for a binary point-mass orbit one finds that (Peters et al. 1963):

$$\dot{\mathcal{P}}/\mathcal{P} = -\frac{3}{2} E_{\mathrm{rad}}^{-1} dE_{\mathrm{rad}}/dt , \tag{4.7.112}$$

i.e. the secular loss of Noetherian energy of the orbit agrees with the energy emitted to infinity in form of gravitational radiation. Hence, $\mathbf{a}_5$ might be called "gravitational radiation reaction acceleration". The quadrupole flux formula has been subject to intensive debates in the past (see e.g. Ehlers et al. 1976, Rosenblum 1978, Cooperstock et al. 1979, Thorne 1980, Walker et al. 1980, Anderson 1980, Rosenblum 1981, Damour 1983, Schäfer 1983, Westpfahl et al. 1987); for further details and references the reader is referred to the Cargèse lectures 1986 by Damour (1987b).

5. Geodesy

5.1 Post-Newtonian Gravimetry

Motivated by the work of Kenneth Nordtvedt, Jr. on the strong equivalence principle Clifford Will (1971,1981) formulated a post-Newtonian theory of gravimetric measurements in the frame of his multi-parameter formalism (parameters $\gamma, \beta, \xi, \alpha_1, \alpha_2, \alpha_3, \zeta_1, \zeta_2, \zeta_3$ and ζ_4 (2.2.18)) with an accuracy of $10^{-9}\, g$. In this theory the post-Newtonian *oscillations* of gravimeter readings, according to Will, are essentially given by:

$$\Delta g/g = \alpha\, \Delta G_L/G_L \ , \tag{5.1.1}$$

where α denotes the usual gravimetric factor and G_L (the locally measured gravitational constant) is given by:

$$
\begin{aligned}
G_L = &\, 1 - [4\beta - \gamma - 3 - \zeta_2 - \xi(3 + I/mr^2)]U_{\text{ext}} \\
&- \tfrac{1}{2}[\alpha_1 - \alpha_3 - \alpha_2(1 - I/mr^2)]w_\oplus^2 \\
&- \tfrac{1}{2}\alpha_2(1 - 3I/mr^2)(\mathbf{w}_\oplus \cdot \hat{\mathbf{e}})^2 + \xi(1 - 3I/mr^2)U_{\text{ext}}^{jk}\,\hat{e}^j\hat{e}^k \ ,
\end{aligned}
\tag{5.1.2}
$$

where I, m and r denote the spherical moment of inertia, mass and radius of the Earth, $\hat{\mathbf{e}}$ is a (Euclidean) unit vector, pointing from the test mass of the gravimeter to the center of the Earth (in usual PN coordinates) and

$$U_{\text{ext}}^{ij} = \sum_{a \neq \oplus} m_a(x_\oplus^i - x_a^i)(x_\oplus^j - x_a^j)/r_{\oplus a}^3 \quad ; \quad U_{\text{ext}} = U_{\text{ext}}^{jj} \ .$$

$w_\oplus$ denotes the velocity of the Earth w.r.t. some preferred frame in the universe, e.g. given by the cosmic microwave background. The first post-Newtonian term in (5.1.2) is related to the Nordtvedt effect (cf. Section 4.7). Due to the eccentricity of the Earth's orbit about the Sun this term leads to an annual variation of g with amplitude of $\sim 10^{-10}\, g$. The remaining terms in (5.1.2), containing the *preferred frame parameters* $\alpha_1, \alpha_2, \alpha_3$ and the *Whitehead parameter* ξ, lead to PN modifications of the diurnal and semi-diurnal tidal amplitudes and the

Earth's spherical moment of inertia. Gravimeter data and data on variations of the length of day (l.o.d.) then yield upper limits for various combinations $(\alpha_2, \xi, \alpha_3 + 2\alpha_2/3 - \alpha_1)$ of PPN parameters. This is described in Will (1981) in detail.

In the following the post-Newtonian theory (parameters: β, γ) of gravimetric measurements will be given with a precision of $10^{-11} g$.

As already discussed in Section 3.3, as observable of gravimetric measurements we will consider the 4-acceleration a the gravimeter's test mass experiences because it is kept "at rest" w.r.t. certain parts of the gravimeter. If one considers the Earth only then the condition to keep the gravimeter's test mass "at rest" is easy to formulate. To a good approximation the metric field in that case is stationary and the test mass is simply at rest in the PN coordinate system. If the influence of other celestial bodies in the solar system is taken into account such a condition is not so easy to formulate. Will, who was interested in the "radial acceleration" of the test body only, formulates such a condition as:

$$\frac{dr_p}{d\tau} = \frac{dr_p}{dt} = 0 \quad ; \quad \frac{d^2 r_p}{d\tau^2} = \frac{d^2 r_p}{dt^2} = 0 \ , \tag{5.1.3}$$

where r_p denotes the proper distance between the gravimeter's test mass and the geocenter. r_p might be defined by one half of the proper time a light ray needs to travel from the test mass through a transparent model Earth and back.[†] This can suitably be calculated by taking the gravitational fields of all celestial bodies in the solar system *apart from the Earth* into account. In *geocentric* post-Newtonian coordinates one finds:

$$r_p = r_{\oplus g} \left(1 - \frac{\mathbf{v}_g \cdot \hat{\mathbf{e}}}{c} + \gamma \sum_a \frac{m_a}{c^2 r_{\oplus a}} \right) \ , \tag{5.1.4}$$

where $\hat{\mathbf{e}} = -\mathbf{x}_g/r_g$ and the sum extends over all massive bodies of the solar system apart from the Earth. The index g refers to the gravimeter's test mass and

$$r_{ab} = \left[\sum_i (x_a^i - x_b^i)^2 \right]^{1/2} \ .$$

The condition (5.1.3) then leads to the relations

$$\mathbf{v}_g \cdot \hat{\mathbf{e}} = \mathcal{O}(\epsilon^3) \tag{5.1.5}$$

$$\frac{d\mathbf{v}_g}{dt} \cdot \hat{\mathbf{e}} = \frac{v_g^2}{r_{\oplus g}} + \mathcal{O}(\epsilon^3) \ , \tag{5.1.6}$$

[†] Here, the differential effect from the gravitational time delay due to the solar gravitational field is negligibly small (cf. Section 5.3.1).

where (5.1.6) represents nothing but the Euclidean projection of the Newtonian equation

$$\frac{dv_g}{dt} = -\nabla U_{\text{centr}} \quad ; \quad U_{\text{centr}} = \frac{1}{2}\Omega^2(x^2 + y^2) \tag{5.1.7}$$

onto the "radial unit vector" ê. In the absence of a more detailed relativistic theory of the motion of the test mass we will use (5.1.7) to determine the magnitude η_G of the acceleration of the test mass. We will assume the geocenter to follow a geodesic, whereas the test mass experiences the 4-acceleration a. Hence ($\tau = \tau_g$):

$$u^\mu_{\oplus;\nu}u^\nu_\oplus = 0 \tag{5.1.8}$$

$$u^\mu_{g;\nu}u^\nu_g = a^\mu \tag{5.1.9}$$

and

$$u^\mu_g a_\mu = 0 \ . \tag{5.1.10}$$

Equation (5.1.9) leads to

$$\frac{d^2t}{d\tau^2} + \Gamma^0_{\nu\lambda}\frac{dx^\nu}{d\tau}\frac{dx^\lambda}{d\tau} = a^0 \tag{5.1.11}$$

and

$$\frac{dv^i_g}{dt} + \Gamma^i_{\mu\nu}(x_g)\frac{v^\mu_g}{c}\frac{v^\nu_g}{c} - \Gamma^0_{\mu\nu}(x_g)\frac{v^i_g}{c}\frac{v^\mu_g}{c}\frac{v^\nu_g}{c} = \left(\frac{d\tau}{dt}\right)^2\left(a^i - \frac{v^i_g}{c}a^0\right) \ . \tag{5.1.12}$$

From (5.1.8) we find

$$\Gamma^i_{00}(x_\oplus) = 0 \tag{5.1.13}$$

and (5.1.10) yields ($a^\mu = (a^0, \mathbf{a})$):

$$a^0 = \frac{\mathbf{v}_g}{c}\mathbf{a} \ . \tag{5.1.14}$$

Therefore,

$$a^i - \frac{v^i_g}{c}\left(\frac{\mathbf{v}_g \cdot \mathbf{a}}{c}\right) = \left(\frac{dt}{d\tau}\right)^2\left(\frac{dv^i_g}{dt} + \Gamma^i_{\mu\nu}(x_g)v^\mu_g v^\nu_g - \Gamma^0_{\mu\nu}(x_g)\frac{v^i_g}{c}v^\mu_g v^\nu_g\right)$$

$$\simeq \left(\frac{dt}{d\tau}\right)^2\left(\frac{dv^i_g}{dt} + c^2\left[\Gamma^i_{00}(x_g) - \Gamma^i_{00}(x_\oplus)\right]\right) \ . \tag{5.1.15}$$

To sufficient accuracy ($m_a = GM_a$):

$$\left(\frac{d\tau}{dt}\right)^2 = 1 - 2\frac{U_\oplus(x_g)}{c^2} - 2\sum_a\frac{m_a}{c^2 r_{\oplus a}} \ . \tag{5.1.16}$$

Inserting the relation (5.1.7) for dv_g/dt we finally obtain the following expression for η_G:

$$\eta_G^2 = (\mathbf{a}, \mathbf{a}) = \left[-\nabla \overline{U} + (\text{Newt. tidal acceleration}) \right.$$
$$\left. -\nabla U_\oplus^{\text{prop}} (3 - 4\beta + \gamma) \sum_a \frac{m_a}{c^2 r_{\oplus a}} \right]^2 , \qquad (5.1.17)$$

where

$$(\text{Newt. tidal acceleration}) = \sum_a m_a \left(\frac{\mathbf{x}_g - \mathbf{x}_a}{r_{ag}^3} - \frac{\mathbf{x}_\oplus - \mathbf{x}_a}{r_{a\oplus}^3} \right) \qquad (5.1.18)$$

and

$$\overline{U} = U_\oplus^{\text{prop}} + U_{\text{centr}}^{\text{prop}} - \frac{(2\beta + \gamma - 2)}{2} \left(\frac{U_\oplus^{\text{prop}}}{c} \right)^2 . \qquad (5.1.19)$$

Here, the index 'prop' indicates that the corresponding terms have to be expressed by means of proper variables and conserved masses. Since (cf. A.1)

$$U(t, \mathbf{x}) = G \int \frac{\rho'}{|\mathbf{x} - \mathbf{x}'|} d^3 x' = \sum_a \frac{m_a}{r_a} - \frac{1}{2} \Phi_1 - 3\gamma \Phi_2 = \hat{U} - \frac{1}{2} \Phi_1 - 3\gamma \Phi_3$$

with

$$\Phi_1 = \sum_a \frac{m_a}{r_a} \left(\frac{v_a}{c} \right)^2 \quad ; \quad \Phi_2 = \sum_a \frac{m_a}{r_a} \sum_{b \neq a} \frac{m_b}{c^2 r_{ab}}$$

for a system of point particles, $U_\oplus^{\text{prop}}$ is determined by

$$\nabla \hat{U}_\oplus = \nabla U_\oplus^{\text{prop}} \left(1 + 2\gamma \sum_a \frac{m_a}{c^2 r_{\oplus a}} \right) , \qquad (5.1.20)$$

where the mass variable in $\hat{U}_\oplus$ is conserved to post-Newtonian order. The last term in (5.1.20) arises because $\hat{U} \propto r^{-2}$. Furthermore,

$$\nabla U_{\text{centr}} = \Omega^2(x, y, 0) = \left(\frac{d\tau}{dt} \right)^2 \Omega'^2(x, y, 0)$$

$$= \left(\frac{d\tau}{dt} \right)^2 \left(1 - \gamma \sum_a \frac{m_a}{c^2 r_{\oplus a}} \right) \Omega'^2(x_p, y_p, 0)$$

$$\equiv \left(\frac{d\tau}{dt} \right)^2 \left(1 - \gamma \sum_a \frac{m_a}{c^2 r_{\oplus a}} \right) \nabla U_{\text{centr}}^{\text{prop}} , \qquad (5.1.21)$$

where the angular velocity $\Omega' = d\varphi/d\tau$ is measured w.r.t. proper time.

In (5.1.17) the last term is again the Nordtvedt term that already appeared in the expression for G_L. The middle term describes the Newtonian tidal acceleration that is not modified by relativistic terms. Such modifications would appear at the level of $10^{-15} g$. If modifications of time and length scales due to the gravitational field of the Sun are taken into account the first term just describes the post-Newtonian acceleration of the test mass in the gravity field of the Earth.

5.2 Realisation of Time Scales

The post-Newtonian theory of the relation between the readings τ of atomic clocks and a suitable time coordinate t has been extensively discussed in Section 3.2. It essentially amounts to an analysis of the relation (3.2.2)

$$\frac{d\tau}{dt} = 1 - \frac{U}{c^2} - \frac{1}{2}\left(\frac{\mathbf{v}}{c}\right)^2 + \mathcal{O}(1/c^4) \ .$$

Here, we would like to discuss three examples of this relation with direct application in modern geodesy.

5.2.1 Synchronisation of Earthbound Clocks by means of Portable Clocks

Expression (3.3.1) for the metric in *co-rotating coordinates* shows that the proper time τ of an earthbound clock is related to the geocentric coordinate time t by

$$c^2 d\tau^2 \simeq \left(1 - 2\frac{U}{c^2} - \frac{\Omega^2(X^2 + Y^2)}{c^2}\right)(c\,dt)^2 - 2\,(\mathbf{\Omega} \times \mathbf{X})\,d\mathbf{X}\,dt - (d\mathbf{X})^2 \ . \quad (5.2.1)$$

If the static oblateness of the Earth is taken into account U is given by:

$$U(\mathbf{X}) \simeq \frac{GM_\oplus}{R}\left(1 - \left(\frac{a_1}{R}\right)^2 J_2\,P_2(\cos\theta)\right) \ , \quad (5.2.2)$$

where $a_1 = 6.378\,14 \times 10^8$ cm denotes the equatorial radius of the Earth and $J_2 = 1.083 \times 10^{-3}$ the static oblateness. We now want to express $U(\mathbf{X})$ in terms of geographic latitude ψ and height h above the geoid. If

$$f \equiv 1 - \frac{a_2}{a_1} \quad (5.2.3)$$

denotes the geometrical flattening of the Earth and a_2 the polar radius of the Earth, then

$$R \simeq a_1 - a_1 f \sin^2 \psi + h \ , \quad (5.2.4)$$

where

$$R \simeq a_1(1 - f \sin^2 \psi)$$

approximately describes the geoid and

$$\cos^2 \theta = \sin^2 \psi(1 - 4f \cos^2 \psi) \simeq \sin^2 \psi \ .$$

Together with the Newtonian *geopotential*

$$\begin{aligned}
U_{\text{geo}}(\mathbf{X}) &= U(\mathbf{X}) + \frac{1}{2}\Omega^2\,(X^2 + Y^2) \\
&\simeq \frac{GM_\oplus}{R}\left(1 + \frac{1}{2}\left(\frac{a_1}{R}\right)^2 J_2(1 - 3\sin^2 \psi)\right) \\
&\quad - \frac{1}{2}\Omega^2 a_1^2(1 - \sin^2 \psi)
\end{aligned} \quad (5.2.5)$$

one finds

$$g(\psi) \equiv - \left. \frac{\partial U_{\text{geo}}}{\partial R} \right|_{R \,=\, a_1(1-f\sin^2\psi)}$$

$$= \frac{GM_\oplus}{a_1^2} + \frac{3GM_\oplus J_2}{2a_1^2} - \Omega^2 a_1 + \sin^2\psi \left(\frac{2GM_\oplus}{a_1^2}f - \frac{9GM_\oplus J_2}{2a_1^2} + \Omega^2 a_1 \right)$$

$$\simeq (9.780\,27 + 0.051\,92 \sin^2\psi)\,10^2\,\text{cm/s}^2 \tag{5.2.6}$$

the latitude dependent gravity acceleration. Hence, in the co-rotating system we can write g_{00} as:

$$g_{00} \simeq 1 - \frac{2}{c^2}U_{\text{geo}}(\mathbf{X}) \simeq 1 + \frac{2}{c^2}(-U_{\text{geo}}^0 + g(\psi)\,h) \;, \tag{5.2.7}$$

if $U_{\text{geo}}^0 = U_{\text{geo}}|_{h=0}$ denotes the geopotential on the geoid. If the time is measured in TAI seconds we can absorb the constant U_{geo}^0 term in the coordinate time t and (5.2.1) reads

$$c^2 d\tau^2 \simeq (1 + 2g(\psi)\,h/c^2)\,(c\,dt)^2 - 2(\mathbf{\Omega} \times \mathbf{X})\,d\mathbf{X}\,dt - (d\mathbf{X})^2 \;. \tag{5.2.8}$$

Hence, on the geoid ($h = 0$) the measured proper time agrees with the *renormalized* geocentric coordinate time. If a clock has a velocity $\mathbf{v}$ w.r.t. the Earth's surface then ($d\mathbf{X} = \mathbf{v}\,dt$):

$$\Delta t \simeq \int d\tau \left[1 - g(\psi)\frac{h}{c^2} + \frac{1}{2}\left(\frac{\mathbf{v}}{c}\right)^2 + \frac{(\mathbf{\Omega} \times \mathbf{X})\cdot\mathbf{v}}{c^2} \right] \tag{5.2.9}$$

with

$$(\mathbf{\Omega} \times \mathbf{X})\frac{\mathbf{v}}{c} = \Omega\,a_1 \cos\psi\,\frac{v_E}{c^2} \;, \tag{5.2.10}$$

where v_E denotes the eastward component of $\mathbf{v}$.

Clocks in the vicinity of the Earth can now be "synchronized" in the following sense:† if we move a portable clock from one earthbound clock to another slowly ($(v/c)^2 \ll 1$) and e.g. at a low height above the geoid, then the elapsed proper time interval during the journey should be related to the elapsed (renormalized) coordinate time difference according to (5.2.8):

$$\Delta t \simeq \Delta\tau + \frac{\Omega\,a_1}{c^2}\int d\tau\,v_e \cos\psi \;. \tag{5.2.11}$$

Conversely, Δt determines the time that an atomic clock on the geoid ($\tau^* = t$) should indicate. We call this procedure: coordinate-time (or asymptotic) synchronization of clocks.

---

† This synchronization is clearly *not* in the sense of Einstein which, according to Section 3.1, would not be possible on the rotating Earth.

In a slow transport along the meridian circles the time indicated by the transportable clocks needs no correction at all ($\Delta t = \Delta \tau$); for a corresponding transport along the parallels we get:

$$\Delta t \simeq \Delta \tau + \frac{\Omega\, a_1}{c^2} L_E \cos \psi \ , \tag{5.2.12}$$

if L_E denotes the distance covered in the eastward direction. If e.g. we move a clock slowly and at a low height along the equator once around the Earth then its reading will differ by about 207 ns from that of a stationary clock.

5.2.2 Comparison of Clocks: Station Clock − Satellite Clock

According to (5.2.1) and (5.2.8) the rate τ of an earthbound clock is related with the geocentric coordinate time t by

$$\tau \simeq \left(1 - \frac{U^0_{\text{geo}}}{c^2} + \frac{g(\psi)\, h}{c^2}\right) t + C_1 \ . \tag{5.2.13}$$

A similar relation for the rate of a clock aboard a satellite is obtained with ($\mu = GM_\oplus$):

$$d\tau_S \simeq \left[1 - \frac{1}{c^2}\left(\frac{\mu}{R} + \frac{1}{2}v_S^2\right)\right] dt = \left[1 - \frac{2\mu}{c^2 R}\left(\frac{1}{R} - \frac{1}{4a}\right)\right] dt \tag{5.2.14}$$

due to the energy integral in the Newtonian Kepler problem

$$v_S^2 = \mu\left(\frac{2}{R} - \frac{1}{a}\right) \ ,$$

where a denotes the semi-major axis of the satellite orbit. With $R = a(1 - e \cos E)$ we can integrate (5.2.13) and obtain (cf. (3.2.14); $\sqrt{\mu a}\, e \sin E = X_S \cdot \dot{X}_S$):

$$\tau_S = \left(1 - \frac{3}{2}\frac{\mu}{c^2 a}\right) t - \frac{2}{c^2}\sqrt{\mu a}\, e \sin E + C_2 \ . \tag{5.2.15}$$

Elimination of t relates τ_S with τ according to

$$\tau_S = \tau + \frac{1}{c^2}\left(U^0_{\text{geo}} - g(\psi)\, h - \frac{3}{2}\frac{\mu}{a}\right)\tau - \frac{2}{c^2}\sqrt{\mu a}\, e \sin E + C \ , \tag{5.2.16}$$

where the constant C can be determined from the time the two clocks were synchronized.

At time $\tau_S(e_E)$ a signal should be emitted from the satellite to the station, arriving there at local proper time $\tau(e_A)$. From (5.2.14) we obtain the elapsed coordinate time intervall Δt that has elapsed between emission (e_E) and absorption (e_A):

$$\Delta t = t_A - t_E = \Delta \tau + \frac{1}{c^2} \left(U^0_{\text{geo}} - g(\psi) h - \frac{3}{2} \frac{\mu}{a} \right) \Delta \tau$$

$$- \frac{2}{c^2} \sqrt{\mu a} \, e \sin E + \text{const.} \qquad (5.2.17)$$

with

$$\Delta \tau = \tau(e_A) - \tau(e_E) \ .$$

In the satellite clocks of the *Global Positioning System* (GPS) the nominal frequency of the satellite clocks have been reduced before launch to compensate for the constant term in (5.2.16). Then only the small periodic relativistic correction due to the eccentricity of the satellite's orbit has to be taken into account.

5.2.3 Doppler Measurements in the Vicinity of the Earth

Let an emitter (E) with velocity $\mathbf{v}_E$ in the gravitational field of the Earth emit some electromagnetic signal at position $\mathbf{x}_E$ that is received by a receiver (A) (position $\mathbf{x}_A$, velocity $\mathbf{v}_A$). Let the frequency of the signal as measured by $E(A)$ be $f_E(f_A)$, and we seek the relation between the two frequencies f_E and f_A. In the post-Newtonian metric, 4-velocity and coordinate induced spatial triad of an observer at rest in the coordinate system are to a sufficient accuracy given by:

$$\overline{e}_{(0)} = \left(1 + \frac{U}{c^2}\right) \frac{\partial}{\partial t} \quad ; \quad \overline{e}_{(i)} = \left(1 - \gamma \frac{U}{c^2}\right) \frac{\partial}{\partial x^i} \qquad (5.2.18)$$

and a Lorentz boost yields the 4-velocity of an observer moving with coordinate velocity $\mathbf{v}$ according to (3.5.3; 3.5.4):

$$e_{(0)} = \Lambda_{(0)}{}^{(\beta)} e_{(\beta)} = \gamma' \left(\overline{e}_{(0)} + \frac{v^i}{c} \overline{e}_{(i)} \right)$$

$$\simeq \left[1 + \frac{1}{c^2} \left(U + \frac{1}{2}\mathbf{v}^2\right)\right] \frac{\partial}{\partial t} + \frac{v^i}{c} \frac{\partial}{\partial x^i} \ . \qquad (5.2.19)$$

Let $\boldsymbol{k}$ now denote the wave vector of a photon, then

$$f = -(\boldsymbol{k}, e_{(0)}) \qquad (5.2.20)$$

yields the frequency as measured by the observer. Let $\tilde{k}^\mu = dx^\mu/dt$ denote the tangent vector to a light ray, then from (3.3.28) we see that $\tilde{k}^\mu$ in the spherical central field is given by ($\hat{\mathbf{n}}$ points in the direction of signal propagation):

$$\tilde{k}^\mu = [1, \hat{\mathbf{n}}(1 - (\gamma + 1)U/c^2)] \ , \qquad (5.2.21)$$

where the light deflection in the central field has been neglected. The wave vector of a photon is then given by

$$k^\mu = f_\infty \frac{dx^\mu}{d\lambda} = f_\infty \frac{dx^\mu}{dt} \frac{dt}{d\lambda} = f_\infty (1 + 2U/c^2)\tilde{k}^\mu \ , \qquad (5.2.22)$$

where $dt/d\lambda$ has been obtained from (3.3.18). Therefore,

$$f = -(\mathbf{k}, \mathbf{e}_{(0)}) = f_\infty \left(1 + \frac{U}{c^2} + \frac{1}{2} \left(\frac{\mathbf{v}}{c} \right)^2 - \boldsymbol{\beta}' \cdot \hat{\mathbf{n}} \right) \qquad (5.2.23)$$

and for the desired relation between f_E and f_A one obtains:

$$\frac{f_A}{f_E} = \frac{1 - \boldsymbol{\beta}'_A \cdot \hat{\mathbf{n}} + U_A/c^2 + \mathbf{v}_A^2/2c^2}{1 - \boldsymbol{\beta}'_E \cdot \hat{\mathbf{n}} + U_E/c^2 + \mathbf{v}_E^2/2c^2} \ . \qquad (5.2.24)$$

5.3 Space Methods

5.3.1 Laser Ranging to Satellites and the Moon

After having extensively discussed the motion of artificial satellites and the Moon, and questions related to frequency and time measurements, we would like to investigate the influence of relativistic effects on the propagation of electromagnetic signals in more detail. The propagation of such signals has already been treated in Section 3.3.2 in the approximation of geometrical optics (light-rays or null-geodesics). From the expression (3.3.29) for the PN trajectory of a light ray in the spherical field of a central body on finds that the elapsed coordinate time interval during which the ray travels from $\mathbf{x}_0$ to $\mathbf{x}$ is given by:

$$t - t_0 = \frac{1}{c} \left[(\mathbf{x} - \mathbf{x}_0) \cdot \hat{\mathbf{n}} + \Delta(t, t_0) \right] \qquad (5.3.1)$$

with

$$\Delta(t, t_0) = (\gamma + 1) \frac{GM}{c^2} \ln \left(\frac{r_N + \mathbf{x}_N \cdot \hat{\mathbf{n}}}{r_0 + \mathbf{x}_0 \cdot \hat{\mathbf{n}}} \right) \ . \qquad (5.3.2)$$

Just as the formal representation of the orbit, this relativistic "time delay correction" depends upon the choice of the coordinates. Expression (5.3.2) is valid in our standard PN-coordinate system (spatially *isotropic* coordinates). If, according to (4.2.2), we switch to a different radial coordinate, then the right hand side of (5.3.2) has to be supplemented by the term

$$\alpha \frac{GM}{c^2} \left(\frac{\mathbf{x}_0}{r_0} - \frac{\mathbf{x}_N}{r_N} \right) \hat{\mathbf{n}} \ . \qquad (5.3.3)$$

In the literature this "light time equation" (5.3.2) is often found in a different form. With

$$\mathbf{d} = \mathbf{x}_0 - \hat{\mathbf{n}}(\mathbf{x}_0 \cdot \hat{\mathbf{n}}) = \mathbf{x} - \hat{\mathbf{n}}(\mathbf{x} \cdot \hat{\mathbf{n}})$$

(**d** connects the gravitating mass with the point of closest approach to the (extended) unperturbed light ray in the Euclidean sense) the argument of the logarithm up to terms of order $\mathcal{O}(\epsilon^2)$ can be written as (Moyer 1981):

$$\frac{r + \mathbf{x} \cdot \hat{\mathbf{n}}}{r_0 + \mathbf{x}_0 \cdot \hat{\mathbf{n}}} = \frac{r + (r^2 - d^2)^{1/2}}{r_0 + (r_0^2 - d^2)^{1/2}} = \frac{r_0 - (r_0^2 - d^2)^{1/2}}{r - (r^2 - d^2)^{1/2}}$$

$$= \frac{r_0 + r + [(r^2 - d^2)^{1/2} - (r_0^2 - d^2)^{1/2}]}{r_0 + r - [(r^2 - d^2)^{1/2} - (r_0^2 - d^2)^{1/2}]}$$

$$= \frac{r_0 + r + \Delta r}{r_0 + r - \Delta r}$$

with $\Delta r \equiv |\mathbf{x} - \mathbf{x}_0|$. The light time equation can then be written as

$$\Delta(t, t_0) = (\gamma + 1) \frac{GM}{c^2} \ln\left(\frac{r_0 + r + \Delta r}{r_0 + r - \Delta r}\right) . \tag{5.3.4}$$

We would now like to discuss the situation of laser ranging to a satellite in more detail. In *barycentric* standard PN coordinates let an earthbound tracking station emit a laser pulse at $(t_e, \mathbf{x}_e)$ which is reflected by some retroreflector at $(t_r, \mathbf{x}_r)$ and returns to the tracking station at $(t_a, \mathbf{x}_a)$. According to (5.3.4):

$$c(t_r - t_e) = r_{er} + (\gamma + 1) \sum_A \frac{GM_A}{c^2} \ln\left(\frac{r_{eA} + r_{rA} + r_{er}}{r_{eA} + r_{rA} - r_{er}}\right) \tag{5.3.5a}$$

and

$$c(t_a - t_r) = r_{ra} + (\gamma + 1) \sum_A \frac{GM_A}{c^2} \ln\left(\frac{r_{aA} + r_{rA} + r_{ra}}{r_{aA} + r_{rA} - r_{ra}}\right) . \tag{5.3.5b}$$

Here, the sum extends over all bodies that contribute to the gravitational time delay correction and

$$r_{er} \equiv |\mathbf{x}_r - \mathbf{x}_e| \qquad \text{etc.}$$

By subtraction one obtains to post-Newtonian order (Martin et al. 1985) (T: tracking station, S: satellite):

$$\frac{c(t_a - t_e)}{2} = \frac{(r_{er} + r_{ra})}{2} + (\gamma + 1) \sum_A \frac{GM_A}{c^2} \ln\left(\frac{r_{AT} + r_{AS} + \Delta r}{r_{AT} + r_{AS} - \Delta r}\right) . \tag{5.3.6}$$

In the sum over A practically only Earth and Sun contribute. We now want to describe the measuring process in a geocentric reference frame. To this end we first remark that the contribution from the Sun to the gravitational time delay to sufficient accuracy can be written as

$$\Delta_\odot(t_a, t_e)/2 = (\gamma + 1) \frac{GM_\odot}{c^2} \ln\left(\frac{r_T + r_S + \Delta r}{r_T + r_S - \Delta r}\right)$$

$$\simeq (\gamma + 1) \frac{GM_\odot}{c^2} \frac{\Delta r}{r_\oplus} \simeq (\gamma + 1) \frac{U_\odot}{c^2} \Delta r \ . \tag{5.3.7}$$

For $(r_{er} + r_{ra})$ we can write:

$$(r_{er} + r_{ra}) = |\mathbf{x}_S(t_r) - \mathbf{x}_T(t_e)| + |\mathbf{x}_S(t_r) - \mathbf{x}_T(t_a)|$$

$$\simeq |\mathbf{b} - \mathbf{v}_\oplus(t_e - t_r)| + |\mathbf{b} - \mathbf{v}_\oplus(t_a - t_r)|$$

$$\simeq 2b - \frac{\mathbf{b} \cdot \mathbf{v}_\oplus}{b}[(t_e - t_r) + (t_a - t_r)]$$

$$+ \frac{1}{2b}\left[\mathbf{v}_\oplus^2 - \left(\frac{\mathbf{b} \cdot \mathbf{v}_\oplus}{b}\right)^2\right][(t_e - t_r)^2 + (t_a - t_r)^2]$$

with

$$\mathbf{b} \equiv \mathbf{x}_S(t_r) - \mathbf{x}_T(t_r) \ .$$

To sufficient approximation :

$$(t_e - t_r) + (t_a - t_r) \simeq -2\frac{\mathbf{b} \cdot \mathbf{v}_\oplus}{c^2}$$

and

$$(t_e - t_r)^2 + (t_a - t_r)^2 \simeq 2\left(\frac{b}{c}\right)^2 \ .$$

Hence, we obtain

$$(r_{er} + r_{ra}) \simeq 2b + b\left[\left(\frac{\mathbf{b} \cdot \mathbf{v}_\oplus}{bc}\right)^2 + \left(\frac{\mathbf{v}_\oplus}{c}\right)^2\right]$$

and (Martin et al. 1985):

$$\frac{c(t_a - t_e)}{2} \simeq b + \frac{b}{2}\left[\left(\frac{\mathbf{b} \cdot \mathbf{v}_\oplus}{bc}\right)^2 + \left(\frac{\mathbf{v}_\oplus}{c}\right)^2\right] + (\gamma + 1)b\frac{U_\odot}{c^2} + \frac{1}{2}\Delta_\oplus(t_a, t_e) \ .$$

$$\tag{5.3.8}$$

If we now switch to geocentric proper variables $(\tau, \mathbf{B})$ according to (3.2.2) and (3.5.13) we obtain the *geocentric* equation corresponding to (5.3.6) in the *barycentric* system:

$$\frac{c(\tau_a - \tau_e)}{2} = B + \frac{1}{2}\Delta_\oplus(\tau_a, \tau_e) \ . \tag{5.3.9}$$

This result indicates that in the co-moving geocentric system all effects from the Sun cancel precisely to the desired accuracy and only the gravitational time delay from the Earth remains. According to the discussion in Section 3.4 it is clear that in the vicinity of the Earth the Sun only gives rise to *tidal effects* that are negligibly small for the gravitational time delay.†

5.3.2 VLBI

Rapid progress in very long baseline radio interferometry (VLBI) (e.g. due to the general use of the Mark III-system) has opened up the possibility of determining astronomical angle distances for small angular separations with an accuracy of about $0.5 \times 10^{-3}\,''$. At this level of accuracy it becomes increasingly important to formulate the whole VLBI measuring process in a relativistic framework, essentially following the radio signals from the sources through the clumpy universe and the solar system into the radio antennas (Fig. 5.1). The cosmological optics part would not be relevant if it were not for proper motions on the apparent celestial sphere. However, geodetic VLBI measurents usually concentrate on certain substructures (hot spots etc.) of quasars that can attain intrinsic velocities comparable with the speed of light. Moreover, apparent proper motions might be amplified by projection effects or gravitational lenses so to attain values of the order of a few $10^{-3}\,''$/y. Here we would like to consider the more "local" relativistic effects on the group delay (Fig. 5.1) of a two-component radio-interferometer system.

As we see from Fig. 5.1 the propagation time difference between the receipts at the two antennas is given in Newton's theory by

$$\Delta\tau = -\frac{\mathbf{b}\cdot\hat{\mathbf{n}}}{c} , \qquad (5.3.10)$$

where $\mathbf{b} = \mathbf{x}_1 - \mathbf{x}_2$ denotes the Newtonian base-vector and $\hat{\mathbf{n}}$ again the unit vector in the direction of signal propagation. For a relativistic treatment of the group delay we will start with the coordinate time interval (5.3.1; 5.3.2), valid in barycentric PN coordinates, where a "light-ray" should be emitted at $(t_0, \mathbf{x}_0)$ and should reach antenna i at $(t_i, \mathbf{x}_i)$. Here the contributions to the gravitational time delay from the Sun and Earth (possibly of Jupiter) can simply be added. For $\Delta t \equiv t_2 - t_1$ we get (Finkelstein et al. 1983):

$$\begin{aligned}
\Delta t &\equiv t_2 - t_1 \\
&= \frac{1}{c}|\mathbf{x}_0(t_0) - \mathbf{x}_2(t_2)| - \frac{1}{c}|\mathbf{x}_0(t_0) - \mathbf{x}_1(t_1)| + \frac{1}{c}[\Delta(t_2, t_0) - \Delta(t_1, t_0)] \\
&\simeq [\mathbf{x}_2(t_2) - \mathbf{x}_1(t_1)]\cdot\hat{\mathbf{n}} + \Delta\tau_{\text{grav}}^{\odot} + \Delta\tau_{\text{grav}}^{\oplus}
\end{aligned} \qquad (5.3.11)$$

---

† If we had renormalized the barycentric time t to TDB the κ_G term in (3.5.22) would have to be taken into account; then the difference between τ (on the geoid) and TDB would have been negligibly small.

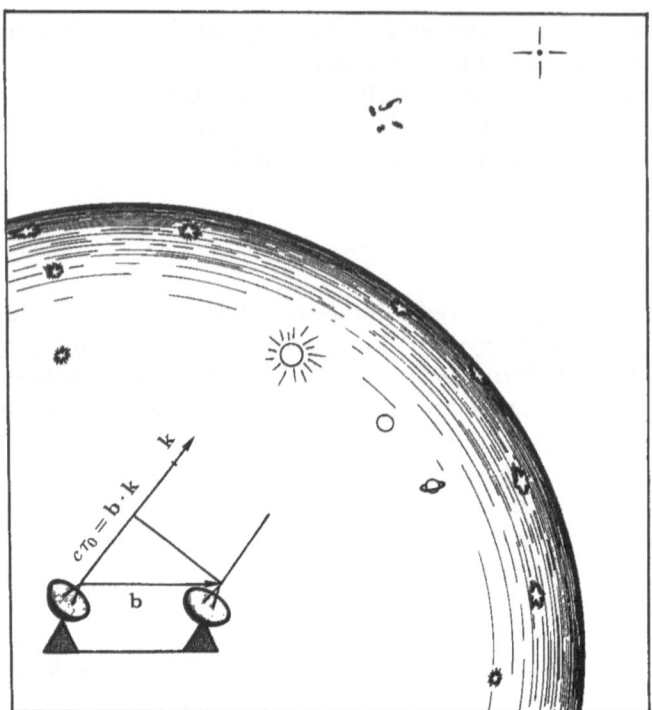

Fig. 5.1.   A two-component radio-interferometer system

with

$$\Delta\tau_{\text{grav}} = (\gamma + 1)\frac{GM}{c^3}\left[\ln\left(\frac{|\mathbf{x}_2| + \mathbf{x}_2 \cdot \hat{\mathbf{n}}_2}{|\mathbf{x}_1| + \mathbf{x}_1 \cdot \hat{\mathbf{n}}_1}\right) + \ln\left(\frac{|\mathbf{x}_0| + \mathbf{x}_0 \cdot \hat{\mathbf{n}}_1}{|\mathbf{x}_0| + \mathbf{x}_0 \cdot \hat{\mathbf{n}}_2}\right)\right] \quad (5.3.12)$$

and

$$\hat{\mathbf{n}}_i = \hat{\mathbf{n}} - \frac{1}{|\mathbf{x}_0|}\left[\hat{\mathbf{n}}(\mathbf{x}_i \cdot \hat{\mathbf{n}}) - \mathbf{x}_i\right]$$

$$- \frac{1}{|\mathbf{x}_0|^2}\left(\mathbf{x}_i(\mathbf{x}_i \cdot \hat{\mathbf{n}}) + \frac{1}{2}\hat{\mathbf{n}}\mathbf{x}_i^2 - \frac{3}{2}\hat{\mathbf{n}}(\mathbf{x}_i \cdot \hat{\mathbf{n}})^2\right) + \cdots. \quad (5.3.13)$$

Here $\hat{\mathbf{n}}$ denotes the Euclidean unit vector from the source to the barycenter. Since

$$\frac{|\mathbf{x}_0| + \mathbf{x}_0 \cdot \hat{\mathbf{n}}_1}{|\mathbf{x}_0| + \mathbf{x}_0 \cdot \hat{\mathbf{n}}_2} \simeq \frac{\mathbf{x}_1^2 - (\mathbf{x}_1 \cdot \hat{\mathbf{n}})^2}{\mathbf{x}_2^2 - (\mathbf{x}_2 \cdot \hat{\mathbf{n}})^2}$$

we obtain

$$\Delta\tau_{\text{grav}} \simeq (\gamma+1)\frac{GM}{c^3}\ln\left(\frac{|\mathbf{x}_1| - \mathbf{x}_1 \cdot \hat{\mathbf{n}}}{|\mathbf{x}_2| - \mathbf{x}_2 \cdot \hat{\mathbf{n}}}\right)$$

$$\simeq (\gamma+1)\frac{GM}{c^3}\ln\left(\frac{1 - \hat{\mathbf{n}}\cdot\hat{\mathbf{s}} + [\hat{\mathbf{s}}\cdot\mathbf{r}_1 - \hat{\mathbf{n}}\cdot\mathbf{r}_1 + r_1^2/2R - (\hat{\mathbf{s}}\cdot\mathbf{r}_1)^2/2R]/R}{1 - \hat{\mathbf{n}}\cdot\hat{\mathbf{s}} + [\hat{\mathbf{s}}\cdot\mathbf{r}_2 - \hat{\mathbf{n}}\cdot\mathbf{r}_2 + r_2^2/2R - (\hat{\mathbf{s}}\cdot\mathbf{r}_2)^2/2R]/R}\right)$$

$$(5.3.14)$$

with

$$\hat{\mathbf{s}} \equiv \frac{\mathbf{x}_\oplus}{|\mathbf{x}_\oplus|} \ . \tag{5.3.15}$$

If $\mathbf{x}_i$ is not parallel to $\hat{\mathbf{n}}$ to a good approximation

$$\Delta\tau_{\text{grav}} \simeq (\gamma+1)\frac{GM}{c^3}\left(\frac{-\mathbf{b}\cdot\hat{\mathbf{n}} + \mathbf{b}\cdot\hat{\mathbf{s}}}{1 - \hat{\mathbf{n}}\cdot\hat{\mathbf{s}}}\right) \ . \tag{5.3.16}$$

For time delay in the field of the Earth ($M = M_\oplus$) $\Delta\tau_{\text{grav}}$ can be written as

$$\Delta\tau_{\text{grav}}^\oplus \simeq (\gamma+1)\frac{GM_\oplus}{c^3}\ln\left(\frac{1 + \sin E_1}{1 + \sin E_2}\right) \ , \tag{5.3.17}$$

where E_i denotes the elevation angle of the source at antenna i.

We now would like to analyze the first term in (5.3.11) in more detail. To this end we first define a barycentric base-vector $\mathbf{b}$ at arrival time of the signal at antenna 1 (this is of course arbitrary):

$$\mathbf{b}(t_1) \equiv \mathbf{x}_1(t_1) - \mathbf{x}_2(t_1) \ , \tag{5.3.18}$$

and a Taylor expansion of $\mathbf{x}_2(t_2)$ about $t = t_2$ yields up to acceleration terms:

$$\Delta t = \Delta\tau_{\text{grav}} - \frac{\mathbf{b}\cdot\hat{\mathbf{n}}}{c}\left[1 + \frac{1}{c}(\mathbf{v}_\oplus + \mathbf{v}_2)\cdot\hat{\mathbf{n}} + \frac{1}{c^2}[(\mathbf{v}_\oplus + \mathbf{v}_2)\cdot\hat{\mathbf{n}}]^2 + \cdots\right] \ , \tag{5.3.19}$$

where we have set $\dot{\mathbf{x}}_2 = \mathbf{v}_\oplus + \mathbf{v}_2$. We now would like to switch from $\mathbf{b}$ to an instantaneous geocentric (proper) base-vector $\mathbf{B}$ (cf. Section 3.5). For $\mathbf{v}_2 = 0$ such a relation (without renormalization) is given by (3.5.22):

$$\mathbf{b} \simeq \left(1 - \gamma\frac{U_\odot}{c^2} - \frac{1}{2}\frac{\mathbf{v}_\oplus \otimes \mathbf{v}_\oplus}{c^2}\right)\mathbf{B} \tag{5.3.20}$$

and, as can be seen by means of a Lorentz transformation, together with the velocity for the Earth's rotation, one obtains

$$\mathbf{b} \simeq \left(1 - \gamma\frac{U_\odot}{c^2} - \frac{1}{2}\frac{\mathbf{v}_\oplus \otimes \mathbf{v}_\oplus}{c^2}\right)\mathbf{B} - \frac{1}{c^2}\mathbf{v}_2(\mathbf{v}_\oplus \cdot \mathbf{B}) \ . \tag{5.3.21}$$

To a good approximation the coordinate time interval Δt can be related to the elapsed proper time difference $\Delta \tau$ by (3.2.2):

$$\frac{\Delta \tau}{\Delta t} \simeq 1 - \frac{U_\odot}{c^2} - \frac{1}{2}\left(\frac{\mathbf{v}_\oplus}{c}\right)^2 \ ,$$

so that we finally obtain (Hellings 1986):

$$\Delta \tau = \Delta \tau_{\text{grav}} - \frac{\mathbf{B} \cdot \hat{\mathbf{n}}}{c}\left[1 + \frac{1}{c}(\mathbf{v}_\oplus + \mathbf{v}_2)\cdot\hat{\mathbf{n}} + \frac{1}{c^2}[(\mathbf{v}_\oplus + \mathbf{v}_2)\cdot\hat{\mathbf{n}}]^2 + \cdots \right.$$

$$\left. - (\gamma + 1)\frac{U_\odot}{c^2} - \frac{1}{2}\left(\frac{\mathbf{v}_\oplus}{c}\right)^2\right]$$

$$- \frac{1}{2c^3}(\mathbf{v}_\oplus \cdot \hat{\mathbf{n}})(\mathbf{v}_\oplus \cdot \mathbf{B}) - \frac{1}{c^3}(\mathbf{v}_2 \cdot \hat{\mathbf{n}})(\mathbf{v}_\oplus \cdot \mathbf{B}) \ . \qquad (5.3.22)$$

This expression is valid for an (asymptotic) barycentric clock synchronization on $t = \text{const.}$ hypersurfaces. If geocentric clock synchronization should be achieved by means of the VLBI process the expression for $\Delta \tau$ has to be augmented by a term of the form:

$$\frac{\mathbf{B} \cdot \mathbf{v}_\oplus}{c^2} \ .$$

A survey of magnitudes of the various relativistic terms in $\Delta \tau$ is given in Table 5.1. The corresponding amplitudes at the limb of Jupiter and Saturn are 1.7 ns and 0.7 ns respectively.

Table 5.1 Magnitudes of Relativistic Effects in $\Delta\tau$ ($b = 6\,000$ km; $\tau_0 \lesssim 20$ ms)

Effect	$\Delta\tau$	Amplitude in ns
Aberration diurnal annual	$\tau_0[\mathbf{v}_2 \cdot \hat{\mathbf{n}}/c + (\mathbf{v}_2 \cdot \hat{\mathbf{n}})^2/c^2]$ $\tau_0[\mathbf{v}_\oplus \cdot \hat{\mathbf{n}}/c + (\mathbf{v}_\oplus \cdot \hat{\mathbf{n}})^2/c^2]$	$40 + 8 \times 10^{-5}$ $2 \times 10^3 + 0.2$
Sun	$\dfrac{r_s^\odot}{c} \ln\left(\dfrac{\|\mathbf{x}_1\| + \mathbf{x}_1 \cdot \hat{\mathbf{n}}}{\|\mathbf{x}_2\| + \mathbf{x}_2 \cdot \hat{\mathbf{n}}}\right)$	limb: 170 $1°$: 45 $\sim 180°$: 0.4
Earth	$\dfrac{r_s^\oplus}{c} \ln\left(\dfrac{1 + \sin E_1}{1 + \sin E_2}\right)$	≤ 0.021
Clock rate	$\dfrac{\tau_0}{c^2}\left[\left(\dfrac{GM_\odot}{R} + \dfrac{1}{2}\mathbf{v}_\oplus^2\right) + \dfrac{GM_\oplus}{R_\oplus}\right]$	$0.3 + 0.014$

Appendix

A.1 PPN-Metric, Christoffel Symbols and Curvature Tensor

Standard post-Newtonian coordinates $(t, \mathbf{x})$; signature $(-+++)$; $c = 1$

PPN Metric (A1.1)

$$g_{00} = -1 + 2U - 2\beta U^2 + 2\Phi$$

$$g_{0i} = -\frac{1}{2}(4\gamma + 3) V_i - \frac{1}{2} W_i$$

$$g_{ij} = (1 + 2\gamma U) \delta_{ij}$$

with:

$$U = G \int \frac{\rho'}{|\mathbf{x} - \mathbf{x}'|} d^3 x'$$

$$\Phi = [(\gamma + 1) \Phi_1 + (3\gamma - 2\beta + 1) \Phi_2 + \Phi_3 + 3\gamma \Phi_4]$$

$$\Phi_1 = G \int \frac{\rho' v'^2}{|\mathbf{x} - \mathbf{x}'|} d^3 x' \quad ; \quad \Phi_2 = G \int \frac{\rho' U'}{|\mathbf{x} - \mathbf{x}'|} d^3 x'$$

$$\Phi_3 = G \int \frac{\rho' \Pi'}{|\mathbf{x} - \mathbf{x}'|} d^3 x' \quad ; \quad \Phi_4 = G \int \frac{p'}{|\mathbf{x} - \mathbf{x}'|} d^3 x'$$

$$V_i = G \int \frac{\rho' v_i'}{|\mathbf{x} - \mathbf{x}'|} d^3 x' \quad ; \quad W_i = G \int \frac{\rho' [\mathbf{v}' \cdot (\mathbf{x} - \mathbf{x}')](x - x')_i}{|\mathbf{x} - \mathbf{x}'|^3} d^3 x'$$

PPN Christoffel Symbols (A1.2)

$$\Gamma^0_{00} = -U_{,0}$$
$$\Gamma^0_{0i} = -U_{,i}$$
$$\Gamma^i_{jk} = \gamma(\delta_{ij}\,U_{,k} + \delta_{ik}\,U_{,j} - \delta_{jk}\,U_{,i})$$
$$\Gamma^i_{0j} = \gamma\delta_{ij}\,U_{,0} - (2\gamma + 2)\,V_{[i,j]}$$
$$\Gamma^i_{00} = -U_{,i} + \frac{\partial}{\partial x^i}[(\beta + \gamma)U^2 - \Phi] - \frac{\partial}{\partial t}\left[\frac{1}{2}(4\gamma + 3)V_i + \frac{1}{2}W_i\right]$$
$$\Gamma^0_{ij} = \gamma\delta_{ij}U_{,0} + \frac{1}{2}(4\gamma + 3)V_{(i,j)} + \frac{1}{2}W_{(i,j)}$$

For the case of an isolated, rotating, stationary and quasi-rigid model Earth the PN potential Φ can be absorbed in U (e.g. Weinberg 1972). For this case also (e.g. Misner et al. 1973):

$$\mathbf{V} = \mathbf{W} = \frac{G\,\mathbf{J}_\oplus \times \mathbf{x}}{2\quad r^3}\ , \tag{A1.3}$$

where $\mathbf{J}_\oplus$ denotes the angular momentum vector of the Earth. The metric coefficients can then be written as:

PPN Metric for an Isolated Earth ($U = U_\oplus$) (A1.4)

$$g_{00} = -1 + 2U - 2\beta U^2$$
$$(g_{0i}) = -(\gamma + 1)G\,\frac{\mathbf{J}_\oplus \times \mathbf{x}}{r^3}$$
$$g_{ij} = (1 + 2\gamma U)\,\delta_{ij}$$

and for the Christoffel symbols one finds:

PPN Christoffel Symbols for an Isolated Earth (A1.5)

$$\Gamma^0_{00} = 0$$
$$\Gamma^0_{0i} = -U_{,i}$$
$$\Gamma^i_{jk} = \gamma(\delta_{ij}\,U_{,k} + \delta_{ik}\,U_{,j} - \delta_{jk}\,U_{,i})$$
$$\Gamma^i_{0j} = -(\gamma + 1)\,(V_{i,j} - V_{j,i})$$
$$\Gamma^i_{00} = -U_{,i} + \frac{\partial}{\partial x^i}[(\beta + \gamma)U^2]$$
$$\Gamma^0_{ij} = (\gamma + 1)\,(V_{i,j} + V_{j,i})$$

For this important case we also want to give the components of the PN Riemann curvature tensor. Here an additional term in Γ^0_{0i}; $[2(\beta - 1)UU_{,i}]$ which is of fourth order in v/c has been taken into account. This term is not needed for the dynamics of test particles.

PPN Curvature Tensor for an Isolated Earth (A1.6)

$$R_{0i0j} = -[U_{,ij} - 2\beta UU_{,ij} - (2\gamma + 2\beta - 1)U_{,i}U_{,j} + \gamma\delta_{ij}U_{,k}U_{,k}]$$

$$R_{0ijk} = (\gamma + 1)\frac{\partial}{\partial x^i}(V_{j,k} - V_{k,j})$$

$$R_{ijkl} = \gamma(\delta_{il}U_{,jk} - \delta_{jl}U_{,ik} + \delta_{kj}U_{,il} - \delta_{ki}U_{,jl})$$

From Table (A1.6) the remaining non-vanishing components of the Riemannian curvature tensor can be obtained from the symmetry relations:

$$R_{\mu\nu\sigma\lambda} = -R_{\nu\mu\sigma\lambda} = -R_{\mu\nu\lambda\sigma} = +R_{\sigma\lambda\mu\nu} \ . \tag{A1.7}$$

If for some problem it is sufficient to consider a purely *spherical* gravitational field it is convenient to employ PPN spherical coordinates. In these coordinates (t, r, θ, ϕ) the metric reads:

$$ds^2 = -\left(1 - \frac{2m}{r} + 2\beta\frac{m^2}{r^2}\right)dt^2 + \left(1 + 2\gamma\frac{m}{r}\right)(dr^2 + r^2(d\theta^2 + \sin^2\theta\, d\phi^2)) \ . \tag{A1.8}$$

PPN Christoffel Symbols for Spherical Field (A1.9)

$$\Gamma^r_{rr} = -\gamma\frac{m}{r^2}$$

$$\Gamma^t_{rt} = \frac{m}{r^2} - 2(\beta - 1)\frac{m^2}{r^3}$$

$$\Gamma^r_{tt} = \frac{m}{r^2} - 2(\beta + \gamma)\frac{m^2}{r^3}$$

$$\Gamma^\theta_{r\theta} = \Gamma^\phi_{r\phi} = \frac{1}{r}\left(1 - \gamma\frac{m}{r}\right)$$

$$\Gamma^r_{\theta\theta} = -r\left(1 - \gamma\frac{m}{r}\right)$$

$$\Gamma^r_{\phi\phi} = -r\sin^2\theta\left(1 - \gamma\frac{m}{r}\right)$$

$$\Gamma^\theta_{\phi\phi} = -\sin\theta\cos\theta$$

$$\Gamma^\phi_{\theta\phi} = \cot\theta$$

PPN Curvature Tensor for Spherical Field (A1.10)

$$R_{trtr} = -\frac{2m}{r^3} + (6\beta + \gamma - 1)\frac{m^2}{r^4}$$

$$R_{t\theta t\theta} = \frac{m}{r}\left(1 - (2\beta + \gamma)\frac{m}{r}\right)$$

$$R_{t\phi t\phi} = \frac{m}{r}\sin^2\theta\left(1 - (2\beta + \gamma)\frac{m}{r}\right)$$

$$R_{r\theta r\theta} = -\gamma\frac{m}{r}$$

$$R_{r\phi r\phi} = -\gamma\sin^2\theta\,\frac{m}{r}$$

$$R_{\theta\phi\theta\phi} = 2\gamma mr\,\sin^2\theta$$

Of great practical importance is the "point mass limit" of the metric and the Christoffel symbols. Here one has to bear in mind that the post-Newtonian conserved mass density is given by (4.7.13)

$$\rho^* = \rho\left(1 + \tfrac{1}{2}\mathbf{v}^2 + 3\gamma U\right) \tag{A1.11}$$

and therefore the PN conserved rest mass $(dM/dt = 0)$ is determined by

$$M = \int \rho^*\, d^3x' \ . \tag{A1.12}$$

Hence, for N point masses $(m = GM;\ a = 1,\ldots,N)$:

$$U(t,\mathbf{x}) = G\int \frac{\rho'}{|\mathbf{x} - \mathbf{x}'|}\, d^3x'$$

$$= G\int \rho^*\frac{(1 - \tfrac{1}{2}\mathbf{v}^2 - 3\gamma U')}{|\mathbf{x} - \mathbf{x}'|}\, d^3x'$$

$$= \sum_a \frac{m_a}{r_a} - \frac{1}{2}\Phi_1 - 3\gamma\,\Phi_2 \ , \tag{A1.13}$$

i.e. in particular

$$U \neq \sum_a \frac{m_a}{r_a}$$

where

$$\mathbf{r}_a = \mathbf{x} - \mathbf{x}_a \quad ; \quad r_a = |\mathbf{x} - \mathbf{x}_a| .$$

By analogy one finds that

$$\Phi_1 = \sum_a \frac{m_a}{r_a} \mathbf{v}_a^2$$

$$\Phi_2 = \sum_a \frac{m_a}{r_a} \sum_{b \neq a} \frac{m_b}{r_{ab}}$$

$$V_i = \sum_a \frac{m_a}{r_a} v_a^i$$

$$W_i = \sum_a \frac{m_a}{r_a^3} (\mathbf{v}_a \cdot \mathbf{r}_a) r_a^i$$

and the PPN metric for a system of point masses reads:

PPN Metric for "point masses" (A1.14)

$$g_{00} = -1 + 2 \sum_a \frac{m_a}{r_a} - 2\beta \left(\sum_a \frac{m_a}{r_a} \right)^2 + (2\gamma + 1) \sum_a \frac{m_a}{r_a} \mathbf{v}_a^2$$

$$- 2(2\beta - 1) \sum_a \frac{m_a}{r_a} \sum_{b \neq a} \frac{m_b}{r_{ab}}$$

$$(\mathbf{g}_{0i}) = -\frac{(4\gamma + 3)}{2} \sum_a \frac{m_a}{r_a} \mathbf{v}_a - \frac{1}{2} \sum_a \frac{m_a}{r_a^3} (\mathbf{v}_a \cdot \mathbf{r}_a) \mathbf{r}_a$$

$$g_{ij} = \left(1 + 2\gamma \sum_a \frac{m_a}{r_a} \right) \delta_{ij}$$

It is instructive to demonstrate how the Christoffel symbols come about for this case, e.g.

$$\Gamma^i_{00} = -U_{,i} + \frac{\partial}{\partial x^i} [(\beta + \gamma)U^2 - \Phi] - \frac{\partial}{\partial t} \left(\frac{1}{2}(4\gamma + 3)V_i + \frac{1}{2} W_i \right)$$

$$= \sum_a \frac{m_a}{r_a^3} r_a^i + 2(\beta + \gamma)U U_{,i} - \left(\gamma + \frac{1}{2} \right) \Phi_{1,i} + (2\beta - 1)\Phi_{2,i}$$

$$- \frac{\partial}{\partial t} \left(\frac{1}{2}(4\gamma + 3)V_i + \frac{1}{2} W_i \right) .$$

Now,

$$\frac{\partial}{\partial t} V_i = \frac{\partial}{\partial t} \left(\sum_a \frac{m_a}{r_a} v_a^i \right) = \sum_a \frac{m_a}{r_a^3} (\mathbf{v_a} \cdot \mathbf{r_a}) v_a^i + \sum_a \frac{m_a}{r_a} \dot{v}_a^i$$

$$= \sum_a \frac{m_a}{r_a^3} (\mathbf{v_a} \cdot \mathbf{r_a}) v_a^i + \sum_a \frac{m_a}{r_a} \sum_{b \neq a} \frac{m_b r_{ab}^i}{r_{ab}^3}$$

and

$$\frac{\partial}{\partial t} W_i = 3 \sum_a \frac{m_a}{r_a^5} (\mathbf{v_a} \cdot \mathbf{r_a})^2 r_a^i - \sum_a \frac{m_a}{r_a^3} \mathbf{v}_a^2 r_a^i - \sum_a \frac{m_a}{r_a^3} (\mathbf{v_a} \cdot \mathbf{r_a}) v_a^i$$

$$+ \sum_a \frac{m_a r_a^i}{r_a^3} \sum_{b \neq a} m_b \left(\frac{\mathbf{r_a} \cdot \mathbf{r_{ab}}}{r_{ab}^3} \right)$$

with

$$\mathbf{r_a} = \mathbf{x} - \mathbf{x_a} \quad ; \quad \mathbf{r_{ab}} = \mathbf{x_b} - \mathbf{x_a} \; .$$

Therefore,

$$\Gamma_{00}^i = \sum_a \frac{m_a}{r_a^3} r_a^i - \sum_a \frac{m_a}{r_a^3} r_a^i 2(\beta + \gamma) \sum_b \frac{m_b}{r_b} + \left(\gamma + \frac{1}{2} \right) \sum_a \frac{m_a}{r_a^3} r_a^i \mathbf{v}_a^2$$

$$- (2\beta - 1) \sum_a \frac{m_a}{r_a^3} r_a^i \sum_{b \neq a} \frac{m_b}{r_{ab}} - \left(\frac{4\gamma + 3}{2} \right) \left[\sum_a \frac{m_a}{r_a^3} (\mathbf{v_a} \cdot \mathbf{r_a}) v_a^i \right.$$

$$\left. + \sum_a \frac{m_a}{r_a} \sum_{b \neq a} \frac{m_b r_{ab}^i}{r_{ab}^3} \right]$$

$$- \frac{1}{2} \left[3 \sum_a \frac{m_a}{r_a^5} (\mathbf{v_a} \cdot \mathbf{r_a})^2 r_a^i - \sum_a \frac{m_a}{r_a^3} \mathbf{v}_a^2 r_a^i \right.$$

$$\left. - \sum_a \frac{m_a}{r_a^3} (\mathbf{v_a} \cdot \mathbf{r_a}) v_a^i + \sum_a \frac{m_a}{r_a^3} r_a^i \sum_{b \neq a} m_b \left(\frac{\mathbf{r_a} \cdot \mathbf{r_{ab}}}{r_{ab}^3} \right) \right] \; ,$$

leading to the expression given in A1.15.

PPN Christoffel Symbols for Point Masses (A1.15)

$$\Gamma^0_{00} = -\sum_a \frac{m_a}{r_a^3} (\mathbf{v}_a \cdot \mathbf{r}_a)$$

$$\Gamma^0_{0i} = \sum_a \frac{m_a}{r_a^3} r_a^i$$

$$\Gamma^i_{jk} = -\gamma \sum_a \frac{m_a}{r_a^3} (\delta_{ij} r_a^k + \delta_{ik} r_a^j - \delta_{jk} r_a^i)$$

$$\Gamma^i_{0j} = \gamma \delta_{ij} \sum_a \frac{m_a}{r_a^3} (\mathbf{v}_a \cdot \mathbf{r}_a) + (\gamma+1) \sum_a \frac{m_a}{r_a^3} (v_a^i r_a^j - v_a^j r_a^i)$$

$$\Gamma^0_{ij} = \left(\gamma+\frac{1}{2}\right) \delta_{ij} \sum_a \frac{m_a}{r_a^3} (\mathbf{v}_a \cdot \mathbf{r}_a) - \frac{3}{2} \sum_a \frac{m_a}{r_a^5} (\mathbf{v}_a \cdot \mathbf{r}_a) r_a^i r_a^j$$

$$- \frac{1}{2}(2\gamma+1) \sum_a \frac{m_a}{r_a^3} (v_a^i r_a^j + v_a^j r_a^i)$$

$$\Gamma^i_{00} = \sum_a \frac{m_a}{r_a^3} r_a^i \left[1 - (2\beta - 1)\sum_{b \neq a} \frac{m_b}{r_{ab}} - 2(\beta+\gamma)\sum_b \frac{m_b}{r_b} + (\gamma+1)\mathbf{v}_a^2\right.$$

$$\left. - \frac{3}{2}\left(\frac{\mathbf{v}_a \cdot \mathbf{r}_a}{r_a}\right)^2 - \frac{1}{2}\sum_{b \neq a}\left(\frac{m_b}{r_{ab}^3}\right)\mathbf{r}_a \cdot \mathbf{r}_{ab}\right]$$

$$- \left(\frac{4\gamma+3}{2}\right) \sum_a \frac{m_a}{r_a} \sum_{b \neq a} \frac{m_b r_{ab}^i}{r_{ab}^3} - (2\gamma+1)\sum_a \frac{m_a}{r_a^3}(\mathbf{v}_a \cdot \mathbf{r}_a) v_a^i$$

A.2 The PPN Two-Body Problem

The Lagrangian (4.1.13) for "spherical bodies" specialized to the two-body problem reads:

$$\mathcal{L} = \mathcal{L}_0 + \mathcal{L}_{PN}/c^2 \tag{A2.1}$$

$$\mathcal{L}_0 = \frac{m_1}{2}\mathbf{v}_1^2 + \frac{m_2}{2}\mathbf{v}_2^2 + \frac{Gm_1m_2}{r}$$

$$\mathcal{L}_{PN} = \frac{1}{8}m_1\mathbf{v}_1^4 + \frac{1}{8}m_2\mathbf{v}_2^4 + \frac{Gm_1m_2}{2r}\left((2\gamma+1)(\mathbf{v}_1^2 + \mathbf{v}_2^2)\right.$$

$$\left. - (4\gamma+3)\mathbf{v}_1 \cdot \mathbf{v}_2 - (\mathbf{v}_1 \cdot \hat{\mathbf{n}})(\mathbf{v}_2 \cdot \hat{\mathbf{n}}) - (2\beta-1)\frac{G(m_1+m_2)}{r}\right)$$

with

$$\hat{\mathbf{n}} = \frac{\mathbf{x}_1 - \mathbf{x}_2}{r} \quad ; \quad r = |\mathbf{x}_1 - \mathbf{x}_2| \ .$$

One finds, that the total momentum **P** of the system can be obtained in the usual manner from $\partial \mathcal{L}/\partial \mathbf{v}_1 + \partial \mathcal{L}/\partial \mathbf{v}_2$ and is given by:

$$\mathbf{P} = m_1 \mathbf{v}_1 + m_2 \mathbf{v}_2 + \frac{1}{2} m_1 \mathbf{v}_1 \frac{v_1^2}{c^2} + \frac{1}{2} m_2 \mathbf{v}_2 \frac{v_2^2}{c^2}$$
$$+ \frac{G m_1 m_2}{2 c^2 r} \Big(2(2\gamma + 1)(\mathbf{v}_1 + \mathbf{v}_2)$$
$$- (4\gamma + 3)(\mathbf{v}_1 + \mathbf{v}_2) - \hat{\mathbf{n}} \left[\hat{\mathbf{n}} \cdot (\mathbf{v}_1 + \mathbf{v}_2) \right] \Big) . \qquad (A2.2)$$

According to the equations of motion the center of mass **X**

$$\mathbf{X} = \frac{(m_1^* \mathbf{x}_1 + m_2^* \mathbf{x}_2)}{(m_1^* + m_2^*)} , \qquad (A2.3)$$

with

$$m_a^* \equiv m_a + \frac{1}{2} m_a \frac{v_a^2}{c^2} - \frac{1}{2} \frac{G m_1 m_2}{r} , \qquad (A2.4)$$

is not accelerated and the center of mass velocity is proportional to **P**. We can then go into a post-Newtonian center of mass system, where $\mathbf{P} = \mathbf{X} = 0$ and

$$\mathbf{x}_1 = \left[\frac{m_2}{m} + \frac{\mu \, \delta m}{2 m^2} \left(\mathbf{v}^2 - \frac{Gm}{r} \right) \right] \mathbf{x} \qquad (A2.5a)$$

$$\mathbf{x}_2 = \left[-\frac{m_1}{m} + \frac{\mu \, \delta m}{2 m^2} \left(\mathbf{v}^2 - \frac{Gm}{r} \right) \right] \mathbf{x} \qquad (A2.5b)$$

with

$$\mathbf{x} = \mathbf{x}_1 - \mathbf{x}_2 \quad ; \quad \mathbf{v} = \mathbf{v}_1 - \mathbf{v}_2 \quad ; \quad m = m_1 + m_2$$
$$\delta m = m_1 - m_2 \quad ; \quad \mu = m_1 m_2 / m .$$

Using (A2.5) the motion of each body is immediately obtained once the *relative motion* is known. For the *relative* motion of the two bodies one then finds:

$$\frac{d\mathbf{v}}{dt} = -\frac{Gm \hat{\mathbf{n}}}{r^2} + \frac{Gm \hat{\mathbf{n}}}{c^2 r^2} \left(\frac{Gm}{r} [2(\beta + \gamma) + 2\nu] - \mathbf{v}^2 (\gamma + 3\nu) + \frac{3}{2} \nu (\hat{\mathbf{n}} \cdot \mathbf{v})^2 \right)$$
$$+ \frac{Gm}{c^2 r^2} \mathbf{v} (\hat{\mathbf{n}} \cdot \mathbf{v})(2\gamma + 2 - 2\nu) \qquad (A2.6)$$

$$\nu \equiv \frac{m_1 m_2}{m^2} = \frac{\mu}{m}$$

and the corresponding Lagrangian reads:

$$\mathcal{L} = \frac{1}{2} \mathbf{v}^2 + \frac{Gm}{r} + \frac{1}{8}(1 - 3\nu) \frac{v^4}{c^2}$$
$$+ \frac{Gm}{2 c^2 r} \left((2\gamma + 1 + \nu)\mathbf{v}^2 + \nu(\hat{\mathbf{n}} \cdot \mathbf{v})^2 - (2\beta - 1)\frac{Gm}{r} \right) . \qquad (A2.7)$$

This Lagrangian is especially useful for the evaluation of the integrals of motion. For the specific post-Newtonian energy $\mathcal{E}$ and the angular momentum $\mathcal{J}$ one finds:

$$\mathcal{E} = \mathbf{v}\frac{\partial \mathcal{L}}{\partial \mathbf{v}} - \mathcal{L} = \frac{1}{2}\mathbf{v}^2 - \frac{m}{r} + \frac{3}{8}(1 - 3\nu)\mathbf{v}^4$$
$$+ \frac{m}{2r}\left((2\gamma + 1 + \nu)\mathbf{v}^2 + \nu(\hat{\mathbf{n}} \cdot \mathbf{v})^2 + (2\beta - 1)\frac{m}{r}\right) \qquad (A2.8)$$

and

$$\mathcal{J} = \left|\mathbf{x} \times \frac{\partial \mathcal{L}}{\partial \mathbf{v}}\right| = |\mathbf{x} \times \mathbf{v}| \left(1 + \frac{1}{2}(1 - 3\nu)\mathbf{v}^2 + (2\gamma + 1 + \nu)\frac{m}{r}\right) . \qquad (A2.9)$$

A.2.1 The Wagoner–Will Representation

As a first step we will follow the route of Wagoner et al. (1976) to derive an expression for the post-Newtonian orbit. The time dependence will then be obtained by analogy from the treatments by Epstein (1977) and Haugan (1985).

In the *Newtonian limit* the solution of (A2.6) is given by

$$\mathbf{x} = r(\cos\phi, \sin\phi, 0)$$

with

$$r = \frac{p}{1 + e\cos(\phi - \omega_0)} \qquad (A2.10)$$

$$r^2\frac{d\phi}{dt} = \sqrt{mp} . \qquad (A2.11)$$

The post-Newtonian solution can then be obtained by setting

$$r^2\frac{d\phi}{dt} = |\mathbf{x} \times \mathbf{v}| = \sqrt{mp}\,(1 + \delta h) \qquad (A2.12)$$

$$\mathbf{v} = \frac{d\mathbf{x}}{dt} = \left(\frac{m}{p}\right)^{1/2}(-\sin\phi, e + \cos\phi, 0) + \mathbf{v}_{PN} . \qquad (A2.13)$$

One finds ($\phi' = \phi - \phi_0$) that

$$r^2\frac{d\phi}{dt} = \sqrt{mp}\left(1 - \frac{me}{p}(2\gamma + 2 - 2\nu)\cos\phi'\right) \qquad (A2.14)$$

and

$$v_{PN}^x = \sqrt{\frac{m^3}{p^3}} \left[-(2\gamma + 2 - \beta)e\phi' + (2\beta + \gamma - \nu)\sin\phi' - \left(\gamma + \frac{21}{8}\nu\right) e^2 \sin\phi' \right.$$
$$\left. + \frac{1}{2}(\beta - 2\nu)e\sin 2\phi' - \frac{\nu}{8}e^2 \sin 3\phi' \right] \qquad (A2.15a)$$

$$v_{PN}^y = \sqrt{\frac{m^3}{p^3}} \left[-(2\beta + \gamma - \nu)\cos\phi' - \left(\gamma + 2 - \frac{31}{8}\nu\right) e^2 \cos\phi' \right.$$
$$\left. - \frac{1}{2}(\beta - 2\nu)e\cos 2\phi' + \frac{\nu}{8}e^2 \cos 3\phi' \right] \qquad (A2.15b)$$

$$v_{PN}^z = 0 . \qquad (A2.15c)$$

An expression for $r(\phi)$ can be obtained by inserting the last two relations into the identity

$$\frac{d}{d\phi}\frac{1}{r} \equiv -\frac{1}{r^2\dot{\phi}} \left(\frac{\mathbf{x} \cdot \mathbf{v}}{r}\right) \qquad (A2.16)$$

and integrating over ϕ. The integration constant is fixed by the requirement that the resulting expression for the post-Newtonian velocity $\mathbf{x}$ is determined by (A2.15). One finds that

$$\frac{p}{r} = 1 + e\cos\phi' + \left(\frac{m}{p}\right) \left[-(2\beta + \gamma - \nu) \right.$$
$$+ \left(\gamma + \frac{9}{4}\nu\right) e^2 + \frac{1}{2}(4\gamma + 4 - \beta - 2\nu)e\cos\phi'$$
$$\left. + (2\gamma + 2 - \beta)e\phi'\sin\phi' - \frac{\nu}{4}e^2 \cos 2\phi' \right] . \qquad (A2.17)$$

From this expression one finds that the *secular* drift in the argument of perigee per revolution is given by

$$\Delta\phi = 2\pi (2\gamma + 2 - \beta)\frac{m}{p} . \qquad (A2.18)$$

This suggests introducing the "true anomaly" η (Epstein 1977, Haugan 1985) with

$$\eta \equiv \left(1 - (2\gamma + 2 - \beta)\frac{m}{p}\right)\phi - \phi_0 \qquad (A2.19)$$

as new angular variable. Then (A2.14) and (A2.17) take the form:

$$r^2 \frac{d\phi}{dt} = \sqrt{mp} \left(1 - \frac{me}{p}(2\gamma + 2 - 2\nu)\cos\eta\right) \qquad (A2.20)$$

and

$$\frac{p}{r} = 1 + e \cos \eta + \left(\frac{m}{p}\right) \left[-(2\beta + \gamma - \nu) + \left(\gamma + \frac{9}{4}\nu\right) e^2 \right.$$
$$\left. + \frac{1}{2}(4\gamma + 4 - \beta - 2\nu)e \cos \eta - \frac{\nu}{4}e^2 \cos 2\eta \right] . \qquad (A2.21)$$

If we furthermore introduce the eccentric anomaly E' instead of the "true anomaly" in the usual manner, i.e.

$$\sin \eta = \frac{(1 - e^2)^{1/2} \sin E'}{1 - e \cos E'} \quad ; \quad \cos \eta = \frac{\cos E' - e}{1 - e \cos E'} , \qquad (A2.22)$$

the last expression can be written as ($p = a(1 - e^2)$):

$$r = a(1 - e \cos E')$$
$$- \frac{m}{c^2(1 - e^2)^2} \left\{ -(2\beta + \gamma) - \frac{(10\gamma + \beta + 12)}{4}e^2 \right.$$
$$+ \frac{1}{2}\gamma e^4 + \left(1 + \frac{17}{4}e^2 + \frac{3}{4}e^4\right)\nu$$
$$+ e \cos E' \left[\left(\frac{8\gamma + 7\beta + 4}{2} + \frac{4 - \beta}{2}e^2\right) - (3 + 5e^2)\nu \right]$$
$$+ e^2 \cos 2E' \left[\left(-\frac{6\gamma + 3\beta + 4}{4} + \frac{1}{2}\gamma e^2\right) + \left(\frac{3}{4} + \frac{5}{4}e^2\right)\nu \right] \right\} . \quad (A2.23)$$

The time dependence of the post-Newtonian relative two-body motion can be expressed by means of a generalized *Kepler equation*. Using (A2.20) and the relations

$$\dot{\phi} = \left(1 + (2\gamma + 2 - \beta)\frac{m}{p}\right)\dot{\eta} \quad ; \quad \dot{\eta} = \frac{(1 - e^2)^{1/2}}{1 - e \cos E'}\dot{E'} \qquad (A2.24)$$

one finds that

$$\left(1 - (2\gamma + 2 - \beta)\frac{m}{p}\right)\sqrt{mp} = \left[1 + \frac{me}{p}(2\gamma + 2 - 2\nu)\left(\frac{\cos E' - e}{1 + e \cos E'}\right)\right] r^2(E')\dot{\eta} ,$$

$$(A2.25)$$

where $r(E')$ is given by (A2.23). Integration of this expression over the time variable t finally gives the desired Kepler equation in the form:

$$t\left(\frac{2\pi}{T_{E'}}\right) + \sigma = E' - ge \sin E' - h \sin 2E' , \qquad (A2.26)$$

where the E'-period $T_{E'}$ is determined by:

$$
T_{E'} = 2\pi \left(\frac{a^3}{m}\right)^{1/2} \left(1 + (2\gamma + 2 - \beta)\frac{m}{p} + \frac{m}{2a(1-e^2)^2}[(8\beta + 4\gamma)\right.
$$
$$
\left. + (6\gamma + \beta + 8)e^2 + (2\gamma + 4)e^4 - (4 + 13e^2 + 7e^4)\nu]\right) \qquad (A2.27)
$$

and

$$
g = 1 + \frac{m}{2a(1-e^2)^2} [8\gamma + 6\beta + 4 - (2\gamma + 3\beta - 4)e^2
$$
$$
- (2\gamma + 4)e^4 - (4 + 11e^2 - 7e^4)\nu] \qquad (A2.28)
$$

$$
h = \frac{e^2 m}{4a(1-e^2)^2} [-(6\gamma + 3\beta + 4) + 2\gamma e^2 + (3 + 5e^2)\nu] \ . \qquad (A2.29)
$$

We finally would like to mention that in this representation the integration constants e and p are related by $\mathcal{E}$ and $\mathcal{J}$ by

$$
\mathcal{E} = -\left(\frac{m}{2p}\right) \left[(1 - e^2) - \left(\frac{m}{4p}\right)\{(8\gamma + 8\beta + 3) - 5\nu\right.
$$
$$
\left. + 2[(4\gamma + 2\beta + 5) - 9\nu]e^2 + 3(1 - 3\nu)e^4\}\right] \qquad (A2.30)
$$
$$
\mathcal{J} = \sqrt{mp} \left[1 + \left(\frac{m}{2p}\right)[(4\gamma + 3 - \nu) + (1 - 3\nu)e^2]\right] \ . \qquad (A2.31)
$$

A.2.2 The Brumberg Representation

The expressions for the post-Newtonian specific energy $\mathcal{E}$ (A2.8) and the magnitude of the angular momentum $\mathcal{J}$ (A2.9) can be written as

$$
\mathcal{E} = \frac{1}{2}(\dot{r}^2 + r^2\dot{\phi}^2) - \frac{m}{r} + \frac{3}{8}(1 - 3\nu)(\dot{r}^2 + r^2\dot{\phi}^2)^2 + \frac{m}{2r}\left((2\gamma + 1 + 2\nu)\dot{r}^2\right.
$$
$$
\left. + (2\gamma + 1 + \nu)r^2\dot{\phi}^2 + (2\beta - 1)\frac{m}{r}\right) \qquad (A2.32)
$$
$$
\mathcal{J} = r^2\dot{\phi} \left(1 + \frac{1}{2}(1 - 3\nu)(\dot{r}^2 + r^2\dot{\phi}^2) + (2\gamma + 1 + \nu)\frac{m}{r}\right) \ . \qquad (A2.33)
$$

This leads us to the first order differential equations of motion:

$$
r^2\dot{\phi} = \mathcal{J} \left(1 + (3\nu - 1)\mathcal{E} + \frac{m}{r}(2\nu - 2\gamma - 2)\right) \qquad (A2.34)
$$

and

$$\dot{r}^2 = -r^2\dot{\phi}^2 + \frac{2m}{r} + 2\mathcal{E} + 2\left[-\frac{3}{2}(1-3\nu)\mathcal{E}^2 + \frac{\nu}{2}\frac{J^2}{r^2}\frac{m}{r}\right.$$

$$\left. -\frac{\mathcal{E}m}{r}(2\gamma+4-7\nu) - \frac{m^2}{r^2}\left(2\gamma+\beta+2-\frac{5}{2}\nu\right)\right] . \quad \text{(A2.35)}$$

Eliminating the $\dot{\phi}^2$ term the last equation can also be written as

$$\dot{r}^2 = A + \frac{2B}{r} + \frac{C}{r^2} + \frac{D}{r^3} \quad \text{(A2.36)}$$

with

$$A = 2\mathcal{E}\left(1+\frac{3}{2}(3\nu-1)\frac{\mathcal{E}}{c^2}\right)$$

$$B = Gm\left(1+(7\nu-2\gamma-4)\frac{\mathcal{E}}{c^2}\right)$$

$$C = -J^2\left(1+2(3\nu-1)\frac{\mathcal{E}}{c^2}\right) + (5\nu-4\gamma-2\beta-4)\frac{G^2m^2}{c^2}$$

$$D = (4\gamma+4-3\nu)\frac{GmJ^2}{c^2} .$$

Using

$$\dot{r}^2 = \left(\frac{dr(\phi(t))}{dt}\right)^2 = J^2\left(\frac{d(1/r)}{d\phi}\right)^2\left[1-2\left((2\gamma+2-2\nu)\frac{m}{r}+(1-3\nu)\mathcal{E}\right)\right]$$

$$\text{(A2.37)}$$

the radial equation can be put into the form:

$$\left(\frac{d(1/r)}{d\phi}\right)^2 = A' + \frac{2B'}{r} + \frac{C'}{r^2} + \frac{D'}{r^3} \quad \text{(A2.38)}$$

with

$$A' = \frac{2\mathcal{E}}{J^2}\left(1+\frac{1}{2}(1-3\nu)\frac{\mathcal{E}}{c^2}\right)$$

$$B' = \frac{Gm}{J^2}\left(1+(2\gamma+2-3\nu)\frac{\mathcal{E}}{c^2}\right)$$

$$C' = -1+(4\gamma+4-2\beta-3\nu)\frac{G^2m^2}{c^2J^2}$$

$$D' = \nu\frac{Gm}{c^2} .$$

We notice that the right hand side of (A2.38) is a third order polynomial in r^{-1}. This suggests writing (A2.38) in the form:

$$\left(\frac{d(1/r)}{d\phi}\right)^2 = \left(\frac{1}{r} - \frac{1}{a(1+e)}\right)\left(\frac{1}{a(1-e)} - \frac{1}{r}\right)\left(C_1 + \frac{C_2}{r}\right) \tag{A2.39}$$

and a comparison of coefficients yields:

$$C_1 = 1 - (4\gamma + 4 - 2\beta - \nu)\frac{m}{a(1-e^2)}$$

$$C_2 = -\nu m \; .$$

From this representation of (A2.39) we see that $r_\pm = a(1 \pm e)$ represents the minimal and maximal values for r, i.e. a and e have the usual meaning as semimajor axis and numerical eccentricity of the post-Newtonian orbit and might be considered as integration constants alternative to $\mathcal{E}$ and $\mathcal{J}$. The relationship between these quantities is given by

$$\mathcal{E} = -\frac{m}{2a}\left[1 - \left(\frac{4\gamma + 3}{4} - \frac{\nu}{4}\right)\frac{m}{a}\right] \tag{A2.40}$$

$$\mathcal{J}^2 = ma(1-e^2)\left[1 + \left(-\gamma - 1 + \nu + \frac{4\gamma + 4 - 2\beta - \nu}{1 - e^2}\right)\frac{m}{a}\right] \; . \tag{A2.41}$$

The solution of (A2.39) is then simply given by

$$r = \frac{a(1-e^2)}{1 + e\cos f} \tag{A2.42}$$

where the 'true anomaly' f obeys the relation

$$\left(\frac{df}{d\phi}\right)^2 = C_1 + \frac{C_2}{r} \; , \tag{A2.43}$$

i.e.

$$\frac{df}{d\phi} = F\left(1 - \frac{\nu}{2}\frac{m}{a(1-e^2)}e\cos f\right) \tag{A2.44}$$

with

$$F = 1 - (2\gamma + 2 - \beta)\frac{m}{a(1-e^2)} \; .$$

Hence,

$$f = F(\phi - \phi_0) - \frac{\nu}{2}\frac{m}{a(1-e^2)}e\sin[F(\phi - \phi_0)] \; , \tag{A2.45}$$

leading us again to (A2.18) for the secular periastron precession. Eliminating $\mathcal{E}$ and $\mathcal{J}$ from (A2.34), we obtain

$$r^2\dot{\phi} = \sqrt{ma(1-e^2)}\left[1 + \left(-\frac{1}{2}(\gamma + 2\nu)\right.\right.$$
$$\left.\left. + \frac{(2\gamma + 2 - \beta - \nu/2)}{(1-e^2)} - 2(\gamma + 1 - \nu)\frac{a}{r}\right)\frac{m}{a}\right] \; ,$$

or together with (A2.44)

$$\sqrt{ma(1-e^2)}\, dt = r^2 df \left[1 + \left(2\gamma + 2 - \frac{3}{2}\nu\right)\frac{m}{r} + \frac{1}{2}(\gamma + 2\nu)\frac{m}{a}\right]. \quad \text{(A2.46)}$$

For a circular orbit we have $r = a$, $e = 0$, and

$$\dot{\phi}^2 \equiv n^2 = \frac{m}{a^3}\left(1 - (2\beta + \gamma - \nu)\frac{m}{a}\right) \quad \text{(A2.47)}$$

therefore defines the mean motion of the post-Newtonian orbit. If we define the eccentric anomaly E and the mean anomaly M via the relations (A2.22) and

$$M = nt + M_0 \;, \quad \text{(A2.48)}$$

the integration of (A2.46) leads to a corresponding Kepler equation in the form

$$M = \left(1 + (2\gamma + 2 - \beta)\frac{m}{a}\right) E - \left[1 + \left(\frac{3}{2}\nu - \beta\right)\frac{m}{a}\right] e \sin E \;. \quad \text{(A2.49)}$$

The *sidereal period* T_ϕ of the orbit during which ϕ increases by 2π is found to be

$$T_\phi = 2\pi\sqrt{\frac{a^3}{m}}\left[1 + \frac{m}{a}\left(\frac{1}{2}(5\gamma + 4 - \nu) - \frac{(2\gamma + 2 - \beta)\sqrt{1-e^2}}{(1 + e\cos f_0)^2}\right)\right]$$

$$= T_f - 2\frac{(2\gamma + 2 - \beta)\sqrt{1-e^2}}{(1 + e\cos f_0)^2}\sqrt{\frac{a^3}{m}} \quad \text{(A2.50)}$$

where the *anomalous period*

$$T_f = 2\pi\sqrt{\frac{a^3}{m}}\left(1 + \frac{1}{2}(5\gamma + 4 - \nu)\frac{m}{a}\right) \quad \text{(A2.51)}$$

indicates the orbital period with respect to axes that precess with the secular perihelion motion.

A.2.3 The Damour-Deruelle Representation

Damour and Deruelle (1985a) have derived another explicit solution of the post-Newtonian two-body problem (for $\beta = \gamma = 1$) that will be discussed in this Section. Let us consider the radial equation (A2.36) for the relative motion first. With the transformation

$$r = \bar{r} + \frac{D}{2C_0} \;,$$

where C_0' is the limit of C when $c^{-1} \to 0$ ($C_0 = -\mathcal{J}^2$) the radial equation takes 'non-relativistic' form:

$$\dot{\overline{r}}^2 = A + \frac{2B}{\overline{r}} + \frac{\overline{C}}{\overline{r}^2} \tag{A2.52}$$

with

$$\overline{C} = C - \frac{BD}{C_0} \; . $$

Hence, $\overline{r}$ can be written as a linear function of $\cos E$, E being an "eccentric anomaly" and the same is true of r. We can therefore write

$$r = a_r \left(1 - e_r \cos E\right) \tag{A2.53}$$

and

$$M \equiv n(t - t_0) = E - e_t \sin E \tag{A2.54}$$

with

$$n = \frac{(-A)^{3/2}}{B} \tag{A2.55a}$$

$$e_t = \left[1 - \frac{A}{B^2} \left(C - \frac{BD}{C_0}\right)\right]^{1/2} \tag{A2.55b}$$

$$a_r = -\frac{B}{A} + \frac{D}{2C_0} \tag{A2.55c}$$

$$e_r = \left(1 + \frac{AD}{2BC_0}\right) e_t \; . \tag{A2.55d}$$

Note that according to (A2.53) a_r and e_r of Damour & Deruelle (1985) coincide with the corresponding parameters a and e of the Brumberg representation ((A2.56a) and (A2.56c) below are inverse to (A2.40) and (A2.41)). The main difference between the post-Newtonian radial motion and the non-relativistic one in the Damour-Deruelle representation is the appearance of two different eccentricities: a 'time-eccentricity' e_t appearing in the Kepler equation (A2.54) and the 'relative radial eccentricity' e_r occuring in (A2.53). In terms of $\mathcal{E}$ and $\mathcal{J}$ the various constants read:

$$a_r = -\frac{Gm}{2\mathcal{E}} \left[1 - \frac{1}{2}(\nu - 4\gamma - 3)\frac{\mathcal{E}}{c^2}\right] \tag{A2.56a}$$

$$e_t = \left\{1 + \frac{2\mathcal{E}}{G^2 m^2}\left[1 + \left(-\frac{7}{2}\nu + 4\gamma + \frac{9}{2}\right)\frac{\mathcal{E}}{c^2}\right]\right.$$
$$\left. \times \left[\mathcal{J}^2 + (-2\nu + 2\beta)\frac{G^2 m^2}{c^2}\right]\right\}^{1/2} \tag{A2.56b}$$

$$e_r = \left\{ 1 + \frac{2\mathcal{E}}{G^2 m^2} \left[1 + \left(\frac{5}{2}\nu - 4\gamma - \frac{7}{2} \right) \frac{\mathcal{E}}{c^2} \right] \right.$$
$$\left. \times \left[\mathcal{J}^2 + (\nu + 2\beta - 4\gamma - 4) \frac{G^2 m^2}{c^2} \right] \right\}^{1/2} \qquad \text{(A2.56c)}$$

$$n = \frac{(-2\mathcal{E})^{3/2}}{Gm} \left(1 - \frac{1}{4}(\nu - 8\gamma - 7)\frac{\mathcal{E}}{c^2} \right) . \qquad \text{(A2.56d)}$$

As consequence of (A2.56a) one can also express n in terms of a_r:

$$n = \left(\frac{m}{a_r^3} \right)^{1/2} \left(1 + \frac{1}{2}(\nu - 5\gamma - 4)\frac{m}{a_r} \right) . \qquad \text{(A2.57)}$$

Note that

$$T_f = \frac{2\pi}{n} = 2\pi \sqrt{\frac{a_r^3}{m}} \left(1 + \frac{1}{2}(5\gamma + 4 - \nu)\frac{m}{a_r} \right) \qquad \text{(A2.58)}$$

denotes the anomalous period, i.e. the time of return to the periastron.

Similarly the post-Newtonian angular motion can be reduced to an auxiliary Keplerian problem. Let us first write (A2.34) as

$$\dot\phi = \frac{H}{r^2} + \frac{I}{r^3} \qquad \text{(A2.59)}$$

with

$$H = \mathcal{J} \left(1 + (3\nu - 1)\frac{\mathcal{E}}{c^2} \right)$$
$$I = (2\nu - 2\gamma - 2)\frac{Gm\mathcal{J}}{c^2} .$$

With

$$r = \tilde{r} + \frac{I}{2H} \qquad \text{(A2.60)}$$

this can simply be written as:

$$\tilde{r}^2 \dot\phi = H . \qquad \text{(A2.61)}$$

Expressing $\tilde{r}$ as

$$\tilde{r} = \tilde{a} \left(1 - \tilde{e} \cos E \right) \qquad \text{(A2.62)}$$

with

$$\tilde{a} = a_r - \frac{I}{2H} \qquad \text{(A2.63)}$$

then together with (A2.54) we obtain

$$d\phi = \frac{H}{n\tilde{a}^2} \frac{1 - e_t \cos E}{(1 - \tilde{e} \cos E)^2} \, dE . \qquad \text{(A2.64)}$$

Choosing a new eccentricity e_ϕ with

$$e_\phi = 2\tilde{e} - e_t \qquad (A2.65)$$

we get

$$\frac{1 - e_t \cos E}{(1 - \tilde{e} \cos E)^2} = \frac{1}{1 - e_\phi \cos E} + \mathcal{O}(\epsilon^4)$$

and

$$d\phi = \frac{H}{n\tilde{a}^2} \frac{dE}{1 - e_\phi \cos E} \cdot \qquad (A2.66)$$

The angular motion is therefore given by

$$\phi - \phi_0 = K f \qquad (A2.67)$$

with

$$\tan\frac{f}{2} = \left(\frac{1 + e_\phi}{1 - e_\phi}\right)^{1/2} \tan\frac{E}{2} \qquad (A2.68)$$

and

$$K = \frac{H}{n\tilde{a}^2(1 - e_\phi^2)^{1/2}} \cdot \qquad (A2.69)$$

In terms of $\mathcal{E}$ and $\mathcal{J}$ the constants e_ϕ and K can be expressed as:

$$e_\phi = (1 - \nu\mathcal{E})e_r = \left(1 + \frac{\mu}{2a_r}\right)e_r$$

$$= \left\{1 + \frac{2\mathcal{E}}{G^2 m^2}\left[1 + \left(\frac{1}{2}\nu - 4\gamma - \frac{7}{2}\right)\frac{\mathcal{E}}{c^2}\right]\left[\mathcal{J}^2 + (2\beta - 4\gamma - 4)\frac{G^2 m^2}{c^2}\right]\right\}^{1/2} \qquad (A2.70)$$

and

$$K = \frac{\mathcal{J}}{[\mathcal{J}^2 + (2\beta - 4\gamma - 4)m^2]^{1/2}} \simeq 1 + (4\gamma - 2\beta + 4)\frac{G^2 m^2}{\mathcal{J}^2 c^2} \cdot \qquad (A2.71)$$

From $r = a_r(1 - e_r \cos E)$ it is obvious that the periastron passages occur for $E = 0, 2\pi, 4\pi, \ldots$ and therefore the periastron advance per revolution is again to post-Newtonian order given by:

$$\Delta\phi = 2\pi(K - 1) = \frac{2\pi(2\gamma + 2 - \beta)Gm}{a_r(1 - e_r^2)c^2} \cdot \qquad (A2.72)$$

The polar equation of the relative orbit is finally found from (A2.53) and (A2.70). Writing (A2.53) as

$$r = \left(\frac{e_r}{e_\phi}\right)a_r(1 - e_\phi \cos E) + a_r\left(1 - \frac{e_r}{e_\phi}\right)$$

and noticing that

$$1 - e_\phi \cos E = \frac{1 - e_\phi^2}{1 + e_\phi \cos f}$$

and

$$a_r \left(1 - \frac{e_r}{e_\phi} \right) = \frac{\mu}{2}$$

we finally get:

$$r = \left(a_r - \frac{G\mu}{2c^2} \right) \frac{1 - e_\phi^2}{1 + e_\phi \cos f} + \frac{G\mu}{2c^2} \qquad (A2.73)$$

for the relative orbit. With

$$f' = f - \frac{1}{2} e_r \frac{\mu}{a_r(1 - e_r^2)} \sin f \qquad (A2.74)$$

this result can also be written as

$$r = \frac{a_r(1 - e_r^2)}{1 + e_r \cos f'} \cdot \qquad (A2.75)$$

A.2.4 The Solution with Osculating Elements

In his monograph Brumberg (1972) treats the *restricted* ($\nu = 0$) two-body problem for a large class of metric theories of gravity by means of parameters σ', β', α' and λ' †. It now turns out that Brumberg's perturbing function is general enough to cover the case of the PPN two-body problem. A comparison of his perturbing function with (A2.6) shows that for

$$\sigma' = \beta + \gamma + \nu$$
$$\beta' = \frac{1}{2}(\gamma + 3\nu)$$
$$\alpha' = \frac{1}{2}\nu \qquad (A2.76)$$
$$\lambda' = \gamma + 1 - \nu$$

Brumberg's results for the osculating elements apply for our PPN two-body problem. The post-Newtonian acceleration $\mathbf{a}_{PN}$ can be written as:

$$\mathbf{a}_{PN} = S\,\mathbf{n} + T\,(\mathbf{k} \times \mathbf{n}) + W\,\mathbf{k}$$

---

† We have added the primes to distinguish the Brumberg parameters from the usual PPN parameters.

with

$$S = \frac{m}{r^2}\left(2(\beta + \gamma + \nu)\frac{m}{r} - (\gamma + 3\nu)\mathbf{v}^2 + (2\gamma + 2 - \frac{1}{2}\nu)\dot{r}^2\right) \quad \text{(A2.77a)}$$

$$T = \frac{m}{r^2}(2\gamma + 2 - 2\nu)\frac{n^2 a^3}{r}e\sin f \qquad \text{(A2.77b)}$$

$$W = 0 \ . \qquad \text{(A2.77c)}$$

The solution of Lagrange's planetary equations is then given by:

$$I = \text{const.} \quad ; \quad \Omega = \text{const.} \qquad \text{(A2.78a)}$$

$$\Delta a = \frac{Gme}{c^2(1 - e^2)^2}\left\{\left[(6\nu - (6\gamma + 4\beta + 4)) + e^2\left(\frac{31}{4}\nu - (4 + 2\gamma)\right)\right]\cos f\right.$$
$$\left. + [4\nu - (2\gamma + 2 + \beta)]e\cos 2f + \frac{\nu}{4}e^2\cos 3f\right\}\Big|_{t_0}^{t} \qquad \text{(A2.78b)}$$

$$\Delta e = \frac{Gm}{c^2 a(1 - e^2)}\left\{\left[(\nu - 2\beta - \gamma) + e^2\left(\frac{47}{8}\nu - 4 - 3\gamma\right)\right]\cos f\right.$$
$$\left. + (2\nu - \gamma - 1 - \frac{1}{2}\beta)e\cos 2f + \frac{\nu}{8}e^2\cos 3f\right\}\Big|_{t_0}^{t} \qquad \text{(A2.78c)}$$

$$\Delta\omega = \frac{Gm}{c^2 a(1 - e^2)}\left\{(2\gamma + 2 - \beta)f + \left[\frac{\nu - \gamma - 2\beta}{e} + \left(\gamma + \frac{21}{8}\nu\right)e\right]\sin f\right.$$
$$\left. + \left(2\nu - \gamma - 1 - \frac{1}{2}\beta\right)\sin 2f + \frac{\nu}{8}e\sin 3f\right\}\Big|_{t_0}^{t} \qquad \text{(A2.78d)}$$

$$\Delta\epsilon = (1 - \sqrt{1 - e^2})\,\Delta\omega + \frac{m}{c^2 a\sqrt{1 - e^2}}[(2\gamma + 4 - 7\nu)\sqrt{1 - e^2}E$$
$$+ (-4\gamma - 4\beta - 4 + 9\nu)f + (4\gamma + 4 - \nu)e\sin f]\big|_{t_0}^{t} \qquad \text{(A2.78e)}$$

$$\int_{t_0}^{t}\Delta n\,dt = \frac{3m}{c^2 a}\left\{-\left(\gamma + 2 - \frac{7}{2}\nu\right)E + \frac{(2\gamma + \beta + 2 - 3\nu)}{\sqrt{1 - e^2}}f - \frac{\nu}{2}\frac{e\sin f}{\sqrt{1 - e^2}}\right.$$
$$+ \left[\frac{\nu}{2}(1 - e^2)\left(\frac{a}{r_0}\right)^3 + \left(-2\gamma - \beta - 2 + \frac{5}{2}\nu\right)\left(\frac{a}{r_0}\right)^2\right.$$
$$\left.\left. + \left(\gamma + 2 - \frac{7}{2}\nu\right)\frac{a}{r_0}\right]M\right\}\Big|_{t_0}^{t} \qquad \text{(A2.78f)}$$

with

$$\Delta M = \Delta\epsilon - \Delta\omega + \int_{t_0}^{t}\Delta n\,dt \ .$$

A.3 On the PPN Hill-Brown Theory

In this part of the Appendix the post-Newtonian Hill-Brown theory for the lunar (barycentric coordinate) motion with PPN parameters β and γ is briefly reviewed. For a much more complete theory in the Einstein case the reader is referred to Brumberg (1958, 1972). The PPN lunar theory is also discussed in Brumberg et al. (1985).

We start with the post-Newtonian equations of motion for a test body a $(m_a \to 0)$ (Moon) in the field of N gravitating bodies of mass m_b $(b = 1, \ldots, n)$, as can be obtained from the Lagrangian (4.1.11):

$$\ddot{\mathbf{x}}_a = \frac{\partial U}{\partial \mathbf{x}_a} - \frac{d}{dt} \frac{\partial U}{\partial \dot{\mathbf{x}}_a} \tag{A3.1}$$

with

$$\begin{aligned}
U &= \mathcal{L}_a - \frac{1}{2}\mathbf{v}_a^2 \\
&= \frac{1}{8c^2}\mathbf{v}_a^4 + \sum_{b \neq a} \frac{m_b}{r_{ab}}\left(1 + \frac{1}{2c^2}(2\gamma+1)(\mathbf{v}_a^2 + \mathbf{v}_b^2) - \frac{1}{2c^2}(4\gamma+3)\mathbf{v}_a \cdot \mathbf{v}_b \right. \\
&\quad \left. - \frac{1}{2c^2}(\mathbf{v}_a \cdot \mathbf{n}_{ab})(\mathbf{v}_b \cdot \mathbf{n}_{ab}) - \frac{1}{c^2}(2\beta-1)\sum_{c \neq a,b}\frac{m_c}{r_{bc}}\right) \\
&\quad - \frac{1}{2c^2}(2\beta-1)\left(\sum_{b \neq a}\frac{m_b}{r_{ab}}\right)^2 .
\end{aligned} \tag{A3.2}$$

We note that the equations of motion (A3.1) remain unchanged by substitutions of the form:

$$U \to U + \frac{dF}{dt} , \tag{A3.3}$$

where F is an arbitrary function of coordinates $(t, \mathbf{x}_a)$ but independent of $\dot{\mathbf{x}}_a$. We now want to address the restricted three-body problem of Earth, Moon and Sun. Let us first neglect the Moon and solve for the post-Newtonian two-body problem of Earth and Sun. To this end we will choose the xy-plane of our PN coordinate system in the orbital plane of the two bodies (that is fixed in space due to angular momentum conservation), the origin in the (Newtonian) center of mass and neglect the eccentricity of the Earth's orbit. According to A.2, the motion of Earth and Sun are then given by (1 = Earth, 2 = Sun):

$$\begin{aligned}
x_1 &= -\frac{m_2}{m}a'\cos n't & x_2 &= +\frac{m_1}{m}a'\cos n't \\
y_1 &= -\frac{m_2}{m}a'\sin n't & y_2 &= +\frac{m_1}{m}a'\sin n't \\
z_1 &= 0 & z_2 &= 0 ,
\end{aligned} \tag{A3.4}$$

where $m = m_1 + m_2$ and a' denotes the semi-major axis of the Earth's orbit that is related with the mean motion n' according to (A2.47)

$$n' = \left(\frac{Gm}{a'^3}\right)^{1/2} \left[1 - \frac{1}{2}\left(\frac{Gm}{c^2 a'}\right)\left((2\beta + \gamma) - \frac{m_1 m_2}{m^2}\right)\right] \ . \tag{A3.5}$$

Inserting (A3.4) into (A3.2) and adding a dF/dt term with

$$F = \frac{1}{2}\frac{Gm_1 m_2}{c^2 m} n'a'(-x\sin n't + y\cos n't)\left(\frac{1}{\rho_1} - \frac{1}{\rho_2}\right) \ , \tag{A3.6}$$

where

$$\rho_b = |\mathbf{x}_a - \mathbf{x}_b| \equiv |\mathbf{x} - \mathbf{x}_b|$$

yields a potential U in the form:

$$\begin{aligned}
U = {}& G\left(\frac{m_1}{\rho_1} + \frac{m_2}{\rho_2}\right) + \frac{1}{c^2}\left\{\frac{1}{8}(\dot{x}^2 + \dot{y}^2 + \dot{z}^2)^2\right. \\
& + \frac{1}{2}\beta_3 G\left(\frac{m_1}{\rho_1} + \frac{m_2}{\rho_2}\right)(\dot{x}^2 + \dot{y}^2 + \dot{z}^2) \\
& - \frac{1}{2}\beta_1 G^2\left(\frac{m_1^2}{\rho_1^2} + \frac{m_2^2}{\rho_2^2}\right) + \frac{Gm_1 m_2}{m}\left[n'a'\left(\left(\alpha_4\dot{x} - \frac{1}{2}n'y\right)\sin n't\right.\right. \\
& + \left(\alpha_4\dot{y} - \frac{1}{2}n'x\right)\cos n't\right)\left(\frac{1}{\rho_1} - \frac{1}{\rho_2}\right) \\
& - \frac{1}{2}\frac{n'^2 a'^2}{m}(-x\sin n't + y\cos n't)^2\left(\frac{m_2}{\rho_1^3} + \frac{m_1}{\rho_2^3}\right) \\
& \left.\left.+ n'^2 a'^2\left(-\beta_1\frac{a'}{\rho_1\rho_2} + \frac{\alpha_1 m_2 - \beta_2 m_1}{2m\rho_1} + \frac{\alpha_1 m_1 - \beta_2 m_2}{2m\rho_2}\right)\right]\right\} \ . \tag{A3.7}
\end{aligned}$$

The constants β_3, β_1 etc. are combinations of β and γ, that in Einstein's theory ($\beta = \gamma = 1$) take the numerical value of their index. The precise values are listed at the end of this Section. Here, and in the following, we will suppress the index a for the lunar orbit. For the determination of the lunar orbit in the frame of the Hill-Brown formalism we also want to indicate the equations of motion for the Moon in geocentric coordinates where the solar gravitational field will be expanded in terms of multipoles.

A.3.1   Geocentric EIH Equations
for the Restricted Three-Body Problem

First, we transform to coordinates (t, ξ, η, ζ) such that the two massive bodies (Earth and Sun) lie on the new ξ-axis. With

$$x = \xi\cos n't - \eta\sin n't \quad ; \quad \xi_1 = -\frac{m_2}{m}a' \quad ; \quad \xi_2 = \frac{m_1}{m}a'$$

$$y = \xi\sin n't + \eta\cos n't \quad ; \quad \eta_1 = 0 \quad\quad ; \quad \eta_2 = 0 \tag{A3.8}$$

$$z = \zeta \quad\quad\quad\quad\quad\quad\quad\quad ; \quad \zeta_1 = 0 \quad\quad ; \quad \zeta_2 = 0$$

the equations of motion can be written as:

$$\ddot{\xi} - 2n'\dot{\eta} = \frac{\partial W}{\partial \xi} - \frac{d}{dt}\frac{\partial W}{\partial \dot{\xi}}$$

$$\ddot{\eta} + 2n'\dot{\xi} = \frac{\partial W}{\partial \eta} - \frac{d}{dt}\frac{\partial W}{\partial \dot{\eta}} \qquad (A3.9)$$

$$\ddot{\zeta} = \frac{\partial W}{\partial \zeta} - \frac{d}{dt}\frac{\partial W}{\partial \dot{\zeta}} \ ,$$

with

$$
\begin{aligned}
W =\ & \frac{1}{2}n'^2(\xi^2 + \eta^2) + U \\
=\ & \frac{1}{2}n'^2(\xi^2 + \eta^2) + G\left(\frac{m_1}{\rho_1} + \frac{m_2}{\rho_2}\right) \\
& + \frac{1}{c^2}\left\{\frac{1}{8}\left(\dot{\xi}^2 + \dot{\eta}^2 + \dot{\zeta}^2 + 2n'(\xi\dot{\eta} - \eta\dot{\xi}) + n'^2(\xi^2 + \eta^2)\right)^2\right. \\
& + \frac{1}{2}\beta_3 G\left(\frac{m_1}{\rho_1} + \frac{m_2}{\rho_2}\right)\left(\dot{\xi}^2 + \dot{\eta}^2 + \dot{\zeta}^2 + 2n'(\xi\dot{\eta} - \eta\dot{\xi}) + n'^2(\xi^2 + \eta^2)\right) \\
& - \frac{1}{2}\beta_1 G^2\left(\frac{m_1^2}{\rho_1^2} + \frac{m_2^2}{\rho_2^2}\right) \\
& + \frac{Gm_1 m_2}{m}\left[n'a'(\alpha_4\dot{\eta} + \frac{1}{2}\alpha_7 n'\xi)\left(\frac{1}{\rho_1} - \frac{1}{\rho_2}\right) - \frac{n'^2 a'^2}{2m}n^2\left(\frac{m_2}{\rho_1^3} + \frac{m_1}{\rho_2^3}\right)\right. \\
& \left.\left.+ n'^2 a'^2\left(-\beta_1\frac{a'}{\rho_1\rho_2} + \frac{\alpha_1 m_2 - \beta_2 m_1}{2m\rho_1} + \frac{\alpha_1 m_1 - \beta_2 m_2}{2m\rho_2}\right)\right]\right\} \ . \quad (A3.10)
\end{aligned}
$$

We now want to switch to geocentric coordinates (t, x', y', z'), where the primes on the coordinates will be omitted. With

$$x = \xi + \frac{m_2}{m}a' \ ; \quad y = \eta \ ; \quad z = \zeta$$

the equations of motion read:

$$\ddot{x} - 2n'\dot{y} = \frac{\partial W}{\partial x} - \frac{d}{dt}\frac{\partial W}{\partial \dot{x}}$$

$$\ddot{y} + 2n'\dot{x} = \frac{\partial W}{\partial y} - \frac{d}{dt}\frac{\partial W}{\partial \dot{y}} \qquad (A3.11)$$

$$\ddot{z} = \frac{\partial W}{\partial z} - \frac{d}{dt}\frac{\partial W}{\partial \dot{z}}$$

and

$$W = W_0 + \frac{1}{8c^2}\left[\dot{x}^2 + \dot{y}^2 + \dot{z}^2 + 2n'(x\dot{y} - y\dot{x}) - 2n'a'\frac{m_2}{m}\dot{y}\right.$$

$$\left. + n'^2\left(x^2 + y^2 - 2\frac{m_2}{m}a'x + \frac{m_2^2}{m^2}a'^2\right)\right]^2$$

$$+ \frac{Gm_1}{c^2\rho_1}\left(-\frac{m_2^2\epsilon_3 + m_1m_2\beta_2}{2m^2}n'^2a'^2 + \frac{1}{2}\frac{m_2}{m}n'^2a'x + \frac{1}{2}\beta_3 n'^2(x^2 + y^2)\right.$$

$$\left. + \frac{m_2}{m}n'a'\dot{y} + \beta_3 n'(x\dot{y} - y\dot{x}) + \frac{1}{2}\beta_3(\dot{x}^2 + \dot{y}^2 + \dot{z}^2)\right)$$

$$+ \frac{Gm_2}{c^2\rho_2}\left(\frac{m_2^2\beta_3 + m_1m_2\alpha_5 + m_1^2\alpha_1}{2m^2}n'^2a'^2 - \frac{2\beta_3 m_2 + \alpha_7 m_1}{2m}n'^2a'x\right.$$

$$+ \frac{1}{2}\beta_3 n'^2(x^2 + y^2)$$

$$\left. - \frac{\beta_3 m_2 + \alpha_4 m_1}{m}n'a'\dot{y} + \beta_3 n'(x\dot{y} - y\dot{x}) + \frac{1}{2}\beta_3(\dot{x}^2 + \dot{y}^2 + \dot{z}^2)\right)$$

$$- \beta_1 \frac{G^2 m_1 m_2}{c^2\rho_1\rho_2} - \frac{1}{2}\beta_1\frac{G^2}{c^2}\left(\frac{m_1^2}{\rho_1^2} + \frac{m_2^2}{\rho_2^2}\right)$$

$$- \frac{Gm_1 m_2}{2c^2 m^2}n'^2a'^2\left(\frac{m_2}{\rho_1^3} + \frac{m_1}{\rho_2^3}\right)y^2 \,, \tag{A3.12}$$

where

$$W_0 = \frac{1}{2}n'^2(x^2 + y^2) + G\left(\frac{m_1}{\rho_1} + \frac{m_2}{\rho_2}\right) - \frac{m_2}{m}n'^2a'x + \cdots \,. \tag{A3.13}$$

We now want to neglect $m_1(M_\oplus)$ w.r.t. $m_2(M_\odot)$ and expand the solar potential in terms of multipoles:

$$\frac{Gm_2}{\rho_2} \simeq n_0'^2 a'^2 + n_0'^2 a'x + \frac{1}{2}n_0'^2(2x^2 - y^2 - z^2)$$

$$+ \frac{n_0'^2}{a'}\left(x^3 - \frac{3}{2}x(y^2 + z^2)\right) + \cdots$$

$$= (1 + \alpha_3\sigma)\left[n'^2 a'x + \frac{1}{2}n'^2(2x^2 - y^2 - z^2)\right.$$

$$\left. + \frac{n'^2}{a'}\left(x^3 - \frac{3}{2}x(y^2 + z^2)\right) + \cdots\right] \,. \tag{A3.14}$$

with

$$\sigma = \frac{Gm_2}{c^2 a'} = \frac{GM_\odot}{c^2 a'} \simeq 10^{-8} \,.$$

For the potential W we then find that

$$
W = W_0 + \sigma \left\{ \left[-\frac{(\epsilon_3 + 2\beta_2)}{2} \frac{Gm_1}{r} - \frac{1}{2} \frac{Gm_1}{r^3} y^2 \right. \right.
$$

$$
+ \frac{(2\beta_3 + 1)}{4} \dot{x}^2 + \frac{(2\beta_3 + 3)}{4} \dot{y}^2 + \frac{(2\beta_3 + 1)}{4} \dot{z}^2
$$

$$
\left. + (\beta_3 + 2) n' x \dot{y} + \frac{1}{4} n'^2 ((4\alpha_3 - 6\beta_1 + 3) x^2 - (2\alpha_3 + \beta_3 - 2\beta_1) z^2) \right]
$$

$$
+ \frac{1}{n' a'} \left[\left(\dot{y} - \frac{(2\beta_1 - 1)}{2} n' x \right) \frac{Gm_1}{r} - \left(\frac{1}{2} \dot{y} - \frac{(\beta_3 - 1)}{2} n' x \right) (\dot{x}^2 + \dot{y}^2 + \dot{z}^2) \right.
$$

$$
\left. - n' \dot{y}(x\dot{y} - y\dot{x}) - \frac{1}{2} n'^2 \dot{y}((4 - \beta_3) x^2 - \beta_3 z^2) \right]
$$

$$
+ \frac{1}{n'^2 a'^2} \left[\frac{1}{2} \beta_3 (\dot{x}^2 + \dot{y}^2 + \dot{z}^2) \frac{Gm_1}{r} - \frac{1}{2} \beta_1 \frac{G^2 m_1^2}{r^2} + \frac{1}{8} (\dot{x}^2 + \dot{y}^2 + \dot{z}^2)^2 \right.
$$

$$
\left. \left. + \beta_3 n'(x\dot{y} - y\dot{x}) \frac{Gm_1}{r} + \frac{1}{2} n'(x\dot{y} - y\dot{x})(\dot{x}^2 + \dot{y}^2 + \dot{z}^2) \right] \right\} \qquad (A3.15)
$$

with $r \equiv \rho_1$. Eliminating the relativistic acceleration terms by means of the Newtonian equations of motion, as, for example, in

$$
-\frac{d}{dt} \frac{\partial W}{\partial \dot{x}} = -\frac{1}{2} \alpha_7 \sigma \ddot{x} = -\frac{1}{2} \alpha_7 \sigma \left(2n' \dot{y} - \frac{Gm_1}{r^3} x + 3n'^2 x \right) ,
$$

for $a'^{-1} \to 0$ the equations of motions can also be written as ($M = M_\oplus$):

$$
\ddot{x} - 2n' \dot{y} + GM \frac{x}{r^3} - 3n'^2 x = \sigma \left[-2\gamma n' \dot{y} - 2\delta_3 n'^2 x + 2\alpha_3 GM \frac{x}{r^3} + \frac{3}{2} GM \frac{xy^2}{r^5} \right]
$$

$$
\ddot{y} + 2n' \dot{x} + GM \frac{y}{r^3} = \sigma \left[+2\alpha_2 n' \dot{x} + 2\alpha_3 GM \frac{y}{r^3} + \frac{3}{2} GM \frac{y^3}{r^5} \right]
$$

$$
\ddot{z} \phantom{- 2n' \dot{y}} + GM \frac{z}{r^3} + n'^2 x = \sigma \left[\phantom{+2\alpha_2 n' \dot{x} aaaaaaaaa} + 2\alpha_3 GM \frac{z}{r^3} + \frac{3}{2} GM \frac{zy^2}{r^5} \right]
$$

$$
(A3.16)
$$

A.3.2 The Geodetic Precession
as derived from the EIH Equations

By transforming (A3.16) back to "nonrotating coordinates" one finds in our case the action of a *constant* relativistic perturbing force that entirely depends upon the relativistic dynamics of the barycenter of the Earth-Moon system about the Sun. To derive such a force we can neglect the relativistic m_1 terms and transform (A3.16) to new coordinates (t, Ξ, H, Z) with

$$
\begin{aligned}
x &= +\Xi \cos n't + H \sin n't \quad ; \quad \Xi = x - x_1 \\
y &= -\Xi \sin n't + H \cos n't \quad ; \quad H = y - y_1 \\
z &= Z \; .
\end{aligned}
\tag{A3.17}
$$

In these "fixed star oriented" coordinates the equations of motion (A3.16) read

$$
\ddot{\Xi} + \frac{Gm_1}{r^3}\Xi - \frac{1}{2}n'^2\Xi - \frac{3}{2}n'^2(\Xi \cos 2n't + H \sin 2n't)
$$

$$
= -2\Omega_{\mathrm{GP}}\dot{H} + \sigma n'[(\dot{H} - \beta_4 n'\Xi)\cos 2n't - (\dot{\Xi} + \beta_4 n'H)\sin 2n't + n'\Xi(1-\beta)]
$$

$$
\ddot{H} + \frac{Gm_1}{r^3}H - \frac{1}{2}n'^2H - \frac{3}{2}n'^2(\Xi \sin 2n't - H \cos 2n't)
$$

$$
= +2\Omega_{\mathrm{GP}}\dot{\Xi} + \sigma n'[(\dot{H} - \beta_4 n'\Xi)\sin 2n't + (\dot{\Xi} + \beta_4 n'H)\cos 2n't + n'H(1-\beta)]
$$

$$
\ddot{Z} + \frac{Gm_1}{r^3}Z + n'^2 Z = 0 \; ,
\tag{A3.18}
$$

where

$$
\Omega_{\mathrm{GP}} = -\frac{(2\gamma + 1)}{2}\sigma n'
$$

is the angular velocity of *geodetic precession* (3.3.87). From this we see the action of a constant relativistic perturbing acceleration R_{GP} with

$$
R_{\mathrm{GP}} = \Omega_{\mathrm{GP}}\left(\Xi\dot{H} - H\dot{\Xi}\right) \; .
\tag{A3.19}
$$

For the lunar orbit the geodetic precession leads to a secular drift of the longitude of perihelion and the node by

$$
\begin{aligned}
\frac{d\varpi}{dt} &= \frac{\tan(I/2)}{na^2\sqrt{1-e^2}}\frac{\partial R}{\partial I} + \frac{\sqrt{1-e^2}}{na^2 e}\frac{\partial R}{\partial e} = -\Omega_{\mathrm{GP}} \\
\frac{d\Omega}{dt} &= \frac{1}{na^2\sqrt{1-e^2}\sin I}\frac{\partial R}{\partial I} = -\Omega_{\mathrm{GP}} \; ,
\end{aligned}
\tag{A3.20}
$$

since R_{GP} can be written as:

$$
R_{\mathrm{GP}} = \Omega_{\mathrm{GP}}\, h_z = \Omega_{\mathrm{GP}}\, h \cos I = \Omega_{\mathrm{GP}}\, na^2\sqrt{1-e^2}\cos I \; .
$$

A.3.3 The PPN Hill-Brown Calculation in the Instantaneous Geocentric System

The Hill-Brown treatment of equations (A3.11) and (A3.15) is best done with a special choice of Hill-Brown variables: they can be chosen so as to eliminate the velocity dependent terms in the relativistic expressions for the variational inequalities. If one introduces new auxiliary coordinates (t, x', y', z') ($\sigma_\epsilon \equiv 1 + \epsilon\sigma$):

$$x' = \sigma_{2\gamma} x \quad ; \quad y' = \sigma_{2\gamma+1/2} y \quad ; \quad z' = \sigma_{2\gamma} z, \qquad \text{(A3.21)}$$

the equations of motion without the parallactic terms read:

$$\ddot{x}' - 2n'\sigma_{-(2\gamma+1)/2}\, \dot{y}' = \sigma_{+4\gamma}\left(\frac{\partial \overline{W}}{\partial x'} - \frac{d}{dt}\frac{\partial \overline{W}}{\partial \dot{x}'}\right)$$

$$\ddot{y}' + 2n'\sigma_{-(2\gamma+1)/2}\, \dot{x}' = \sigma_{+4\gamma}\left(\frac{\partial \overline{W}}{\partial y'} - \frac{d}{dt}\frac{\partial \overline{W}}{\partial \dot{y}'}\right) \qquad \text{(A3.22)}$$

$$\ddot{z}' = \sigma_{+4\gamma}\left(\frac{\partial \overline{W}}{\partial z'} - \frac{d}{dt}\frac{\partial \overline{W}}{\partial \dot{z}'}\right)$$

with a potential $\overline{W}$

$$\overline{W} = \frac{Gm_1}{r} + \frac{1}{2}n'^2(3x^2 - z^2) + \cdots$$
$$- \sigma\left(2\alpha_3\frac{Gm_1}{r} + \delta_3 n'^2 x'^2 + \frac{1}{2}Gm_1\frac{y^2}{r^3}\right), \qquad \text{(A3.23)}$$

that now is independent of $\dot{x}$. Let

$$D \equiv (n - n')(t - t_0)$$

$$\tilde{m} \equiv \frac{n'}{n - n'}\sigma_{-(2\gamma+1)/2} \quad ; \quad \kappa \equiv \frac{Gm_1}{(n - n')^2}\sigma_{+4(\gamma - \beta)} \qquad \text{(A3.24)}$$

and define the Hill-Brown variables u, s by:

$$u \equiv x' + iy' \quad ; \quad s \equiv x' - iy' \, . \qquad \text{(A3.25)}$$

The equations of motion can then be written as:

$$\frac{d^2 u}{dD^2} + 2\tilde{m}i\,\frac{du}{dD} = 2\left(\frac{\partial \Omega}{\partial s} - \frac{d}{dD}\frac{\partial \Omega}{\partial ds/dD}\right)$$

$$\frac{d^2 s}{dD^2} - 2\tilde{m}i\,\frac{ds}{dD} = 2\left(\frac{\partial \Omega}{\partial u} - \frac{d}{dD}\frac{\partial \Omega}{\partial du/dD}\right) \qquad \text{(A3.26)}$$

$$\frac{d^2 z'}{dD^2} = \left(\frac{\partial \Omega}{\partial z'} - \frac{d}{dD}\frac{\partial \Omega}{\partial dz'/dD}\right),$$

with

$$\Omega = \frac{(1+4\gamma\sigma)}{(n-n')^2}\overline{W}$$

$$= \frac{\kappa}{R} + \frac{3}{8}\tilde{m}^2(u+s)^2\sigma_{\gamma_3/3} - \frac{1}{2}\tilde{m}\sigma_{\beta_3}z'^2 + \cdots$$

$$+ \frac{\sigma}{\tilde{m}a'}\left\{\frac{1}{4}i\left(\frac{du}{dD} - \frac{ds}{dD}\right)\left[-\frac{2\kappa}{R} + \frac{du}{dD}\frac{ds}{dD} + \left(\frac{dz'}{dD}\right)^2\right]\right.$$

$$+ \frac{\tilde{m}}{4}(u+s)\left[-(2\beta_1-1)\frac{\kappa}{R} + (\beta_3-2)\frac{du}{dD}\frac{ds}{dD} + (\beta_3-1)\left(\frac{dz'}{dD}\right)^2\right]$$

$$+ \frac{\tilde{m}}{4}\left[u\left(\frac{ds}{dD}\right)^2 + s\left(\frac{du}{dD}\right)^2\right] - \frac{1}{8}\tilde{m}^2i\left[(4-\beta_3)\left(u^2\frac{ds}{dD} - s^2\frac{du}{dD}\right)\right.$$

$$\left.\left. + 2\beta_3\left(\frac{du}{dD} - \frac{ds}{dD}\right)z'^2\right]\right\}$$

$$+ \frac{\sigma}{\tilde{m}^2a'^2}\left\{\frac{1}{2}\beta_3\left[\frac{du}{dD}\frac{ds}{dD} + \left(\frac{dz'}{dD}\right)^2\right]\frac{\kappa}{R} - \frac{1}{2}\beta_1\frac{\kappa^2}{R^2}\right.$$

$$+ \frac{1}{8}\left[\frac{du}{dD}\frac{ds}{dD} + \left(\frac{dz'}{dD}\right)^2\right]^2$$

$$\left.+ \frac{\tilde{m}}{4}i\left(u\frac{ds}{dD} - s\frac{du}{dD}\right)\left[2\beta_3\frac{\kappa}{R} + \frac{du}{dD}\frac{ds}{dD} + \left(\frac{dz'}{dD}\right)^2\right]\right\} + \cdots \text{ (A3.27)}$$

and $R \equiv \sqrt{us + z'^2}$.

We first want to solve for the auxiliary variables (x', y', z') or (u, s, z'). To this end we write:

$$\Omega = \Omega^{(0)} + \sigma\,\Omega^{(1)}$$

with

$$\Omega^{(0)} = \frac{\kappa}{R} + \frac{3}{8}\tilde{m}^2(u+s)^2 - \frac{1}{2}\tilde{m}^2z'^2 + \Omega'$$

and the equations of motion as $(\mathcal{D} = -i\partial/\partial D)$:

$$\mathcal{D}^2u + 2\tilde{m}\,\mathcal{D}u + \frac{3}{2}\tilde{m}^2(u+s) - \frac{\kappa u}{R^3} = -2\frac{\partial\Omega'}{\partial s} - 2\sigma\left(\frac{\partial\Omega^{(1)}}{\partial s} - \mathcal{D}\frac{\partial\Omega^{(1)}}{\partial\mathcal{D}s}\right)$$

$$\mathcal{D}^2s - 2\tilde{m}\,\mathcal{D}s + \frac{3}{2}\tilde{m}^2(u+s) - \frac{\kappa s}{R^3} = -2\frac{\partial\Omega'}{\partial u} - 2\sigma\left(\frac{\partial\Omega^{(1)}}{\partial u} - \mathcal{D}\frac{\partial\Omega^{(1)}}{\partial\mathcal{D}u}\right)$$

$$\mathcal{D}^2z' - \tilde{m}^2z' - \frac{\kappa z'}{R^3} = -\frac{\partial\Omega'}{\partial z'} - 2\sigma\left(\frac{\partial\Omega^{(1)}}{\partial z'} - \mathcal{D}\frac{\partial\Omega^{(1)}}{\partial\mathcal{D}z'}\right) .$$

$$\text{(A3.28)}$$

If we subtract the second equation multiplied by u, from the first multiplied by s, we obtain:

$$\mathcal{D}(u\mathcal{D}s - s\mathcal{D}u - 2\tilde{m}\,us) + \frac{3}{2}\tilde{m}^2(u^2 - s^2) + \tilde{m}^2 S = \sigma\Psi \ . \qquad (A3.29)$$

By analogy with the classical treatment (e.g. Brown 1960) we find the second Hill equation as

$$\mathcal{D}^2(us + z'^2) - \mathcal{D}u\mathcal{D}s - (\mathcal{D}z')^2 - 2\tilde{m}\,(u\mathcal{D}s - s\mathcal{D}u)$$
$$+ \frac{9}{4}\tilde{m}^2(u + s)^2 - 3\tilde{m}^2 z'^2 + \tilde{m}^2 L = C + \sigma\Phi \ , \qquad (A3.30)$$

where C is an integration constant and

$$S = \frac{1}{a'}\left(\frac{15}{8}(u^3 - s^3) + \frac{3}{8}us(u - s)\right) + \cdots$$

$$L = \frac{1}{a'}\left(\frac{5}{2}(u^3 + s^3) + \frac{3}{2}us(u + s)\right) + \cdots$$

$$\Psi = -2\left(u\frac{\partial\Omega^{(1)}}{\partial u} - s\frac{\partial\Omega^{(1)}}{\partial s}\right) + 2\left(u\mathcal{D}\frac{\partial\Omega^{(1)}}{\partial\mathcal{D}u} - s\mathcal{D}\frac{\partial\Omega^{(1)}}{\partial\mathcal{D}s}\right)$$

$$\Phi = -2\left(u\frac{\partial\Omega^{(1)}}{\partial u} + s\frac{\partial\Omega^{(1)}}{\partial s} + z'\frac{\partial\Omega^{(1)}}{\partial z'} + \Omega^{(1)}\right)$$
$$+ 2\mathcal{D}\left(u\frac{\partial\Omega^{(1)}}{\partial\mathcal{D}u} + s\frac{\partial\Omega^{(1)}}{\partial\mathcal{D}s} + z'\frac{\partial\Omega^{(1)}}{\partial\mathcal{D}z'}\right) \ .$$

These equations (A3.29, A3.30) can now be solved in the usual way by means of power series for u and s. Here we only indicate the computation of the variational terms, i.e. in the following we will neglect the solar parallax (a'^{-1}) and the inclination of the lunar orbit (i.e. z'). In this case: $S = L = \Omega' = 0$ and

$$\Psi = -\frac{1}{2}\gamma_3\,\tilde{m}^2\,(u^2 - s^2) \qquad (A3.31a)$$

$$\Phi = -\frac{3}{4}\gamma_3\,\tilde{m}^2\,(u + s)^2 \ . \qquad (A3.31b)$$

If we furthermore neglect the eccentricity of the lunar orbit the Hill-Brown equations can be solved by setting:

$$u = \mathbf{a}\sum_{-\infty}^{+\infty} a_{2k}\,\zeta^{2k+1} \quad ; \quad s = \mathbf{a}\sum_{-\infty}^{+\infty} a_{-2k-2}\,\zeta^{2k+1}$$

with

$$\zeta = \exp(iD) \ .$$

Equations (A3.29) and (A3.30) then take the form:

$$\mathbf{a}^{-2}C - \frac{3}{2}\gamma_3\sigma\tilde{m}^2 \sum_k a_{2k}\left(A_{-2k-2} + a_{2k}\right)$$

$$= \sum_k \left[(2k+1)^2 + 4\tilde{m}(2k+1) + \frac{9}{2}\tilde{m}^2\right] a_{2k}^2 + \frac{9}{2}\tilde{m}^2 \sum_k a_{2k}a_{-2k-2} \quad (A3.32a)$$

$$\sigma\Phi_{2l} = \sum_k \left[4l^2 + (2k+1)(2k+1-2l) + 4\tilde{m}(2k+1-l) + \frac{9}{2}\tilde{m}^2\right] a_{2k}\, a_{-2l+2k}$$

$$+ \frac{9}{4}\tilde{m}^2 \sum_k a_{2k}\left(A_{2l-2k-2} + a_{-2l-2k-2}\right) \quad (A3.32b)$$

$$\sigma\Psi_{2l} = -\sum_k 4l(2k+1-l+\tilde{m})\, a_{2k}\, a_{-2l+2k}$$

$$+ \frac{3}{2}\tilde{m}^2 \sum_k \left(A_{2l-2k-2} - a_{-2l-2k-2}\right), \quad (A3.32c)$$

with

$$\Phi_{2l} = -\frac{3}{4}\gamma_3\,\tilde{m}^2 \sum_k a_{2k}(A_{2l-2k-2} + 2a_{2k-2l} + a_{-2l-2k-2})$$

$$\Psi_{2l} = -\frac{1}{2}\gamma_3\,\tilde{m}^2 \sum_k a_{2k}(A_{2l-2k-2} - a_{-2l-2k-2}).$$

Equation (A3.32a) can be used to determine the integration constant C and (A3.32b) and (A3.32c) together yield:

$$\sum_k \left\{[2l, 2k]\, a_{2k}\, a_{2k-2l} + [2l]\, a_{2k}\, a_{2l-2k-2} + (2l)\, a_{2k}\, a_{-2l-2k-2}\right\} = \sigma\,\chi_{2l}$$

$$(A3.33)$$

with

$$\chi_{2l} = \frac{1}{9\tilde{m}^2}\left\{2\{[2l] + (2l)\}\Phi_{2l} + 3\{[2l] - (2l)\}\Psi_{2l}\right\}$$

and the Hill-Brown brackets:

$$[2l, 2k] = -\frac{k}{l}\,\frac{8l^2 - 2 - 4\tilde{m} + \tilde{m}^2 + 4(k-l)(l-1-\tilde{m})}{8l^2 - 2 - 4\tilde{m} + \tilde{m}^2} \quad (l \neq 0)$$

$$[2l] = -\frac{3}{16}\,\tilde{m}^2 l^2\,\frac{4l^2 - 8l - 2 - 4(l+2)\tilde{m} - 9\tilde{m}^2}{8l^2 - 2 - 4\tilde{m} + \tilde{m}^2}$$

$$(2l) = -\frac{3}{16}\,\tilde{m}^2 l^2\,\frac{20l^2 - 16l + 2 - 4(5l-2)\tilde{m} + 9\tilde{m}^2}{8l^2 - 2 - 4\tilde{m} + \tilde{m}^2}.$$

Now, σ is a small quantity of order 10^{-8} and we can determine χ_{2l} by the "Newtonian expressions" for a_{2k}. To lowest order one finds:

$$\Phi_2 = -\frac{3}{4}\gamma_3\,\tilde{m}^2 + \cdots \quad ; \quad \Psi_2 = -\frac{1}{2}\gamma_3\,\tilde{m}^2 + \cdots \qquad (A3.34a)$$

$$\chi_{+2} = -\gamma_3\left(\frac{1}{16}\tilde{m}^2 + \frac{1}{6}\tilde{m}^3 + \cdots\right) \quad ; \quad \chi_{-2} = \frac{1}{3}\gamma_3\left(\frac{19}{16}\tilde{m}^2 + \frac{5}{3}\tilde{m}^3 + \cdots\right). \qquad (A3.34b)$$

Now, the σ dependence of the coefficients a_{2k} is twofold: i) *explicitly* through the $\sigma\chi$ term and ii) *implicitly* by $\tilde{m} = m\sigma_{-(2\gamma+1)/2}$. If we denote the *explicit* dependence by

$$a_{2k} = \bar{a}_{2k} + \sigma\delta^{(1)}a_{2k}$$

from (A3.33) we obtain the relation

$$\sum_k \left\{ [2l, 2k]\bar{a}_{2k-2l} + [2l, 2k+2l]\bar{a}_{2k} + 2[2l]\bar{a}_{2l-2k-2} \right.$$
$$\left. + 2(2l)\bar{a}_{-2l-2k-2} \right\} \delta^{(1)}a_{2k} = \chi_{2l}\ , \qquad (A3.35)$$

which can be solved by iteration. To lowest order, for example,

$$\delta^{(1)}a_{+2} = -\chi_{+2} \quad ; \quad \delta^{(1)}a_{-2} = -\chi_{-2}\ . \qquad (A3.36)$$

Together with the implicit σ dependence in $\tilde{m}$ one obtains the post-Newtonian corrections to the Newtonian coefficients a_{2k}, like

$$\delta a_{+2} = -\frac{1}{8}\delta_3\,m^2 - \frac{1}{12}\alpha_{21}\,m^3 + \cdots \qquad (A3.37a)$$

$$\delta a_{-2} = \frac{19}{24}\delta_3\,m^2 + \frac{1}{18}\,\alpha_{105}\,m^3 + \cdots\ . \qquad (A3.37b)$$

The post-Newtonian scale factor **a** is determined by (A3.32a):

$$\mathbf{a} = \bar{a}_0 + \sigma\delta a \quad ; \quad \frac{\delta a}{a_0} = \frac{4}{3}(\gamma - \beta) + \frac{1}{3}\beta_3\,m - \frac{1}{9}\alpha_9\,m^2 + \cdots \qquad (A3.38)$$

with

$$\bar{a}_0 = a_0\left(1 - \frac{1}{6}m^2 + \frac{1}{3}m^3 + \cdots\right) \quad ; \quad n_0^2 a_0^3 = GM_\oplus\ .$$

$\bar{a}_0$ simply denotes the semi-major axis of the lunar orbit if the Newtonian tidal forces from the Sun are taken into account. We finally want to switch to instantaneous geocentric coordinates (t, x, y, z). With

$$x + iy = \sigma_{-(8\gamma+1)/4}\,u + \frac{1}{4}\sigma s \equiv \tilde{\mathbf{a}}\sum_k \tilde{a}_{2k}\,\zeta^{2k+1} \qquad (A3.39a)$$

$$x - iy = \sigma_{-(8\gamma+1)/4}\,s + \frac{1}{4}\sigma u \equiv \tilde{\mathbf{a}}\sum_k \tilde{a}_{-2k-2}\,\zeta^{2k+1} \qquad (A3.39b)$$

we then find that

$$\tilde{\mathbf{a}} = \mathbf{a} \left(1 - \tfrac{1}{4}(8\gamma + 1)\sigma + \tfrac{1}{4}\sigma a_{-2}\right) \tag{A3.40a}$$

$$\tilde{a}_{2k} = \left(1 - \tfrac{1}{4}\sigma a_{-2}\right) a_{2k} + \tfrac{1}{4}\sigma a_{-2k-2} \tag{A3.40b}$$

Note, that $\tilde{a}_{2k}$ is *not* of order $m^{|2k|}$ though this was the case for a_{2k}. To lowest order, for example,

$$\delta\tilde{a}_{+2} = -\tfrac{1}{8}\delta_3\, m^2 + \cdots \quad ; \quad \delta\tilde{a}_{-2} = \tfrac{1}{4} + \tfrac{19}{24}\delta_3\, m^2 + \cdots \; . \tag{A3.41}$$

Note the appearance of the m independent $\sigma/4$ term in $\tilde{a}_{-2}$ that leads to the dominant relativistic oscillation term in the instantaneous geocentric system. The origin of that term is easily traced back to the different σ dependence of y' and x' or z', more precisely to the additional $\sigma_{+1/2}$ factor in the definition of y'. This factor is just eliminated by a Lorentz boost to a proper reference frame, where m independent oscillations in distance do not appear. From (A3.38) one furthermore finds that

$$\frac{\delta\tilde{a}}{\tilde{a}_0} = -\tfrac{2}{3}\gamma - \tfrac{4}{3}\beta - \tfrac{1}{4} + \tfrac{1}{3}\beta_3 m - m^2 \left(\tfrac{1}{9}\alpha_9 + \tfrac{19}{64}\right) + \cdots \; . \tag{A3.42}$$

Transformation to polar coordinates is done in the usual way:

$$r\cos(v - n(t - t_0)) = \tilde{\mathbf{a}} \left[1 + (\tilde{a}_{+2} + \tilde{a}_{-2})\cos 2D + \cdots\right]$$
$$r\sin(v - n(t - t_0)) = \tilde{\mathbf{a}} \left[(\tilde{a}_{+2} - \tilde{a}_{-2})\sin 2D + \cdots\right] \; ,$$

leading us to the expressions for the variational terms in coordinate distance and longitude:

$$\frac{r}{a_0} = 1 - m^2 \cos 2D + \cdots$$
$$+ \sigma \left\{ \left[-\tfrac{2}{3}\gamma - \tfrac{4}{3}\beta - \tfrac{1}{4} + \tfrac{1}{3}(2\gamma + 1)m - \left(\tfrac{10}{3}\gamma - 3\beta + \tfrac{109}{96}\right)m^2 + \cdots\right] \right.$$
$$\left. + \left[\tfrac{1}{4} + \left(\tfrac{1}{4} + 2\gamma + 2\beta\right)m^2\right]\cos 2D + \tfrac{7}{32}m^2 \cos 4D + \cdots \right\} \tag{A3.43}$$

$$v = n(t - t_0) + \tfrac{11}{8}m^2 \sin 2D + \cdots$$
$$+ \sigma \left[\left(-\tfrac{1}{4} - \tfrac{11}{12}(2\gamma + \beta)m^2\right)\sin 2D - \tfrac{11}{32}m^2 \sin 4D + \cdots\right] \; . \tag{A3.44}$$

In this Section various combinations of the PPN parameters β and γ have been abbreviated. The precise values for the various abbreviations read:

$$\alpha_1 = 2\gamma - 4\beta + 3 \qquad\qquad \beta_1 = 2\beta - 1 \qquad\qquad \gamma_3 = 2\gamma - 2\beta + 3$$
$$\alpha_2 = \gamma + 1 \qquad\qquad\qquad \beta_2 = 2(2\beta - 1)$$
$$\alpha_3 = \gamma + 2\beta \qquad\qquad\quad\ \beta_3 = 2\gamma + 1 \qquad\qquad \delta_3 = 2\gamma + \beta$$
$$\alpha_4 = 2\gamma + 2 \qquad\qquad\qquad \beta_4 = 2\gamma + \beta + 1$$
$$\alpha_5 = 4\gamma - 4\beta + 5 \qquad\qquad\qquad\qquad\qquad\qquad \epsilon_3 = 4\beta - 1\ .$$
$$\alpha_7 = 4\gamma + 3$$
$$\alpha_9 = 3\,(10\gamma - 9\beta + 2)$$
$$\alpha_{21} = 14\gamma + 4\beta + 3$$
$$\alpha_{105} = 5\,\alpha_{21}$$

A.4 On the PPN Euler Equation

In *harmonic coordinates* the Euler equation of motion for an ideal fluid (energy momentum tensor given by (4.7.9)) in the PPN metric (4.7.4) reads:

$$\frac{\partial}{\partial t}[\rho(1 + \mathbf{v}^2 + 2U + \Pi + p/\rho)\mathbf{v}] + \nabla \cdot [\mathbf{v}\mathbf{v}\rho(1 + \mathbf{v}^2 + 2U + \Pi + p/\rho)]$$

$$= -\nabla[p(1 - 2\gamma U)] + \rho\nabla(U - (\beta + \gamma)U^2 - \psi) - \rho\boldsymbol{\zeta}_{,0}$$

$$+ \rho(\mathbf{v}^2 + 2U + \Pi)\nabla U + \rho\mathbf{v} \times (\nabla \times \boldsymbol{\zeta}) - (5\gamma - 1)\rho\mathbf{v}U_{,0}$$

$$- (2\gamma - 1)p\nabla U + \gamma\rho\mathbf{v}^2\nabla U - (5\gamma - 1)\rho\mathbf{v}(\mathbf{v} \cdot \nabla U) \qquad (A4.1)$$

with

$$\boldsymbol{\zeta} = -2(\gamma + 1)\mathbf{V}$$

and

$$\psi = -\Phi + \frac{1}{2}\chi_{,00}\ .$$

A detailed derivation of (A4.1) for $\beta = \gamma = 1$; $\Pi = 0$ can be found in Weinberg (1972). Using $\chi_{,0i} = V_i - W_i$ and therefore

$$\psi_{,i} = -\Phi_{,i} + \frac{1}{2}(V_i - W_i)_{,0}$$

this can also be written as

$$\frac{\partial}{\partial t}[\rho(1 + \mathbf{v}^2 + 2U + \Pi + p/\rho)v^i] + \frac{\partial}{\partial x^j}[\rho(1 + \mathbf{v}^2 + 2U + \Pi + p/\rho)v^i v^j]$$

$$= -[p(1 - 2\gamma U)]_{,i} + \rho\left(U_{,i} - (\beta + \gamma)U_{,i}^2 + \Phi_{,i} + \frac{(4\gamma + 3)}{2}V_{i,0} + \frac{1}{2}W_{i,0}\right)$$

$$+ \rho(\mathbf{v}^2 + 2U + \Pi)U_{,i} + (2\gamma + 2)\rho v^j(V_{i,j} - V_{j,i}) - (5\gamma - 1)\rho v^i U_{,0}$$

$$- (2\gamma - 1)p\,U_{,i} + \gamma\rho\mathbf{v}^2 U_{,i} - (5\gamma - 1)\rho v^i v^j U_{,j})\ . \qquad (A4.2)$$

Making use of the Newtonian equations of motion

$$\rho_{,0} + (\rho v^i)_{,i} = 0$$

and

$$(\rho v^i)_{,0} + (\rho v^i v^j + p\delta_{ij}) = \rho U_{,i}$$

we obtain to PN order:

$$\frac{\partial}{\partial t}[\rho(1 + \mathbf{v}^2 + 2U + \Pi + p/\rho)v^i] - (2\gamma + 2)\rho V_{i,0} + (5\gamma - 1)\rho v^i U_{,0}$$

$$= \frac{\partial}{\partial t}\left[\rho^*\left(1 + \tfrac{1}{2}\mathbf{v}^2 - U + \Pi + p/\rho^*\right)v^i + (2\gamma + 2)\rho^*(v^i U - V^i)\right]$$

$$- (2\gamma + 2)V^i(\rho v^j)_{,j} + (5\gamma - 1)U[(\rho v^i v^j + p\delta_{ij})_{,j} - \rho U_{,i}]$$

$$= \frac{\partial}{\partial t}\left[\rho^*\left(1 + \tfrac{1}{2}\mathbf{v}^2 - U + \Pi + p/\rho^*\right)v^i + (2\gamma + 2)\rho^*(v^i U - V^i)\right]$$

$$+ (2\gamma + 2)\rho^* v^j V^i_{,j} - (5\gamma - 1)[U_{,j}(\rho v^i v^j + p\delta_{ij}) + \rho^* U U_{,i}]$$

$$- \frac{\partial}{\partial x^j}[(2\gamma + 2)\rho^* v^j V^i - (5\gamma - 1)U(\rho^* v^i v^j + p\delta_{ij})] \ . \tag{A4.3}$$

Using this result (A4.2) can be written as:

$$\frac{\partial}{\partial t}\left[\rho^*\left(1 + \tfrac{1}{2}\mathbf{v}^2 - U + \Pi + p/\rho^*\right)v^i + (2\gamma + 2)\rho^*(v^i U - V^i)\right]$$

$$+ \frac{\partial}{\partial x^j}\left[\rho^*\left(1 + \tfrac{1}{2}\mathbf{v}^2 - U + \Pi + p/\rho^*\right)v^i v^j + (2\gamma + 2)\rho^*(v^i U - V^i)v^j\right]$$

$$+ [1 + (3\gamma - 1)U]p\,\delta_{ij} - \rho U_{,i} - \rho^*(\varphi U_{,i} + \Phi_{,i})$$

$$+ (2\gamma + 2)\rho^* v^j V_{j,i} + \tfrac{1}{2}\rho^*(V_i - W_i)_{,0} = 0 \ . \tag{A4.4}$$

The time derivative in $V_i - W_i$ can be absorbed by the first term since:

$$\rho(V_i - W_i)_{,0} = [\rho(V_i - W_i)]_{,0} + (V_i - W_i)(\rho v^j)_{,j}$$

$$= [\rho(V_i - W_j)]_{,0} + [\rho v^j(V_i - W_i)] - \rho v^j(V_i - W_i)_{,j} \ .$$

Therefore the Euler equation (A4.1) can be written in the "quasi-Newtonian" form (Chandrasekhar et al. 1969, Caporali 1979, 1981):

$$\frac{\partial}{\partial t}\pi^i + \frac{\partial}{\partial x^j}\sigma^{ij} = f^i \tag{A4.5}$$

with

$$\pi^i = \rho^* \left(1 + \tfrac{1}{2}\mathbf{v}^2 - U + \Pi + p/\rho^*\right) v^i + (2\gamma + 2)\rho^*(v^i U - V^i)$$

$$+ \tfrac{1}{2}\rho^*(V^i - W^i)$$

$$\sigma^{ij} = \pi^i v^j + [1 + (3\gamma - 1)U]p\,\delta_{ij}$$

and

$$f^i = \rho\, U_{,i} + \rho^*(\varphi U_{,i} + \Phi_{,i}) - (2\gamma + 2)\rho^* v^j V_{j,i} + \tfrac{1}{2} v^j (V^i - W^i)_{,j} \ .$$

References

Anderson, J. D., 1974, in: *Experimental Gravitation: Proceedings of Course 56 of the International School of Physics "Enrico Fermi"*, ed. by B. Bertotti, 163, Academic, New York

Anderson, J. L., 1980, *Phys. Rev. Lett.* **45**, 1745

Anderson, J. L., Decanio, T. C., 1975, *Gen. Rel. Grav.* **6**, 197

Anderson, J. D., Keesey, M. S. W., Lau, E. L., Standish, E. M., Newhall, XX 1978, *Acta Astronautica* **5**, 43

Anselmo, L., Farinella, P., Milani, A., Nobili, A., 1983, *Astron. Astrophys.* **117**, 3

Aoki, S., 1964, *Astron. J.* **69**, 221

Arifov, L., Kadyev, R., 1968, in: *Abstracts of Fifth Intern. Conf. on Gravitation and Relativity Theory*, Tbilisi University Press; see also: *Astron. Zh.* **45**, 1114

Arnowitt, R., Deser, S., Misner, C., 1960, *Phys. Rev.* **120**, 313

Ashby, N., Bertotti, B., 1984, *Phys. Rev. Lett.* **52**, 485

Ashtekar, A., 1977, in: *Asymptotic Structure of Spacetime*, F.P. Esposito, L. Witten (ed.), Plenum Press, New York

— 1980, in: *General Relativity and Gravitation*, vol. 2, A. Held (ed.), Plenum Press, New York

Baierlein, R., 1967, *Phys. Rev.* **162**, 1275

Barker, B. M., O'Connell, R. F., 1970, *Phys. Rev.* **D2**, 1428

— 1975, *Phys. Rev.* **D12**, 329

— 1976, *Phys. Rev.* **D14**, 861

— 1981, *Phys. Rev.* **D24**, 2332

Barker, B. M., Byrd, G. G., O'Connell, R. F., 1982, *Astrophys. J.* **253**, 309

— 1986, *Astrophys. J.* **305**, 623

Bailey, I., Israel, W., 1980, *Ann. of Phys.* **130**, 188

Bertotti, B., Ciufolini, I., Bender, P., 1987, *Phys. Rev. Lett.* **58**, 1062

Blackwell, D., Petford, A., 1966, *Mon. Not. Roy. Ast. Soc.* **131**, 383

Blackwell, D., Dewhist, D., Ingham, M., 1967, in: *Advances in Astronomy and Astrophysics*, Academic Press, New York

Blanchet, L., Damour, T., 1988, *Phys. Rev.* **D37**, 1410

Blandford, R., Teukolsky, S., 1976, *Astrophys. J.* **205**, 580

Bretagnon, P., 1982, *Astron. Astrophys.* **114**, 278

Brown, E. W., 1960, *An Introductory Treatise on the Lunar Theory*, Dover

Brumberg, V. A., 1958, *Bull. Inst. Theor. Astron.* (USSR) **6**, 733

— 1972, *Relativistic Celestial Mechanics*, Nauka (in Russian)

— 1981a, in: E.Gaposchkin and B.Kolaczek (eds.) *Reference Coordinate Systems for Earth Dynamics*, Reidel, Dordrecht

— 1981b, *Sov. Astron.* **25**, 101

Brumberg, V. A., Finkel'shtein, A., 1979, *Sov. Phys. JETP* **49**, 749

Brumberg, V. A., Ivanova, T. V., 1982, in: *Sun and Planetary System*, eds. W. Fricke and G. Teleki, Reidel, Dordrecht

— 1985, works of the institute for theoretical astrophysics, Nauka, Vol. XIX

Campbell, W., Trumpler, R., 1923a, *Lick Observ. Bull.* **11**, 41

— 1923b, *Publ. Astron. Soc. Pacific* **35**, 158

— 1928, *Lick Observ. Bull.* **13**, 130

Caporali, A., 1979, *An Approximation Method for the Determination of the Motion of Extended Bodies in General Relativity*, Ph.D. Thesis, Univ. of Munich

— 1981, *Nuovo Cimento B* **61**, 181, 205 and 213

Carmeli, M., 1964, *Phys. Lett.* **9**, 132

— 1965, *Nuovo Cimento* **37**, 842

Cartan, É., 1923, *Ann. École Norm. Sup.* **40**, 325

— 1924, *Ann. École Norm. Sup.* **41**, 1

Carter, W. E., Robertson, D. S., MacKay, J. R., 1985, *J. Geophys. Res.* **90**, 4577

Chandrasekhar, S., 1965, *Astrophys. J.* **142**, 1488

Chandrasekhar, S., Nutku, Y., 1969, *Astrophys. J.* **158**, 55

Chandrasekhar, S., Esposito, F., 1970, *Astrophys. J.* **160**, 153

Chant, C., Young, R., 1924, *Publ. Dominion Astron. Obs.* **2**, 275

Chapront-Touzé, M., 1980, *Astron. Astrophys.* **83**, 86

Chapront-Touzé, M., Chapront, J., 1983, *Astron. Astrophys.* **124**, 50

Chazy, J., 1928, 1930, *La Théorie de la Relativité et la Mécanique Céleste*, vols. 1 and 2, Gauthier-Villars, Paris

Ciufolini, I., 1986a, *Phys. Rev. Lett.* **56**, 278

Ciufolini, I., 1986b, *Found. Phys.* **16**, 259

Cooperstock, F., Hobill, D., 1979, *Phys. Rev.* **D20**, 2995

Counselman, C., Kent, S., Knight, C., Shapiro, I., Clark, T., Hinteregger, H., Rogers, A., Whitney, A., 1974, *Phys. Rev. Lett.* **33**, 1621

Damour, T., 1982, *C.R. Acad. Sci. Paris* **294**, (II) 1355

— 1983a, *Phys. Rev. Lett.* **51**, 1019

— 1983b, in: *Gravitational Radiation*, N. Deruelle and T. Piran (eds.), North-Holland, Amsterdam

— 1987a, in: *Three Hundred Years of Gravitation*, ed. by S. W. Hawking and W. Israel, Cambridge University Press, Cambridge

— 1987b, in: *Gravitation in Astrophysics*, ed. by B. Carter and J. B. Hartle, Plenum Press, New York

Damour, T., Deruelle, N., 1981a, *Phys. Lett.* **87A**, 81

— 1981b, *C.R. Acad. Sci. Paris* **293**, (II) 537 and 877

— 1985a, *Ann. Inst. Poincaré* **43**, 107

— 1986, *Ann. Inst. Poincaré* **44**, 263

Damour, T., Schäfer, G., 1985b, *Gen. Rel. Grav.* **17**, 879

— 1987, *C.R. Acad. Sci. Paris* **305**, 839

— 1988, *Higher-Order Relativistic Periastron Advances and Binary Pulsars*, Nuovo Cim. B in press

Davies, M. M., Taylor, J. H., Weisberg, J. M., Backer, D. C., 1985, *Nature* **315**, 547

de Sitter, W., 1916, *Mon. Not. Roy. Ast. Soc.* **57**, 155

Dixon, W. G., 1970a, *Proc. Roy. Soc. London* **A314**, 499

— 1970b, *Proc. Roy. Soc. London* **A319**, 509

— 1973, *Gen. Rel. Grav.* **4**, 199

— 1974, *Proc. Roy. Soc. London* **A277**, 59

— 1976, *Extended Bodies in General Relativity: Their description and motion*, in: *Isolated Gravitating Systems in General Relativity*, Ehlers, J. (ed.), North-Holland, Amsterdam 1979

Dodwell, G., Davidson, C., 1924, *Mon. Not. Roy. Ast. Soc.* **84**, 150

Droste, J., 1916, *Versl. K. Akad. Wet. Amsterdam* **19**, 447

Dyson, F., Eddington, A., Davidson, C., 1920a, *Phil. Trans. Roy. Soc.* **220A**, 291

Dyson, F., Eddington, A., Davidson, C., 1920b, *Mem. Roy. Ast. Soc.* **62**, 291

Eddington, A. S., 1975, *The Mathematical Theory of Relativity*, Chelsea, New York

Eddington, A. S., Clark, G. L., 1938, *Proc. Roy. Soc. (Lond.)* **A166**, 465

Ehlers, J., 1961, *Abh. Akad. Wiss. u. Lit. Mainz*, Math.-Nat. Kl., No 11, 793

— 1973, in: *Relativity, Astrophysics and Cosmology*, ed. by W. Israel, Reidel, Dordrecht

— 1980, *Ann. N.Y. Acad. Sci.* **336**, 279

— 1981, in: *Grundlagenprobleme der modernen Physik*, J.Nitsch et al. (eds.), Bibliogr. Inst., Mannheim

— 1986, in: *Logic, Methodology and Philosophy of Science VII*, B. Marcus et al. (eds.), Elsevier Science Publishers B. V.

Ehlers, J., Rosenblum, A., Goldberg, J., Havas, P., 1976, *Astrophys. J. Lett.* **208**, 77

Ehlers, J., Rudolph, E., 1977, *Gen. Rel. Grav.* **8**, 197

Einstein, A., 1915, *Preuss. Akad. Wiss. Berlin, Sitzber.* pp. 778, 799, 831 and 844

Einstein, A., Infeld, L., Hoffmann, B., 1938, *Ann. Math.* **39**, 65

Epstein, R., 1977, *Astrophys. J.* **216**, 92 erratum **231**, 644

Epstein, R., Shapiro, I., 1980, *Phys. Rev.* **D22**, 2947

Everitt, C. W. F., 1974, in: *Experimental Gravitation: Proc. of Course 56 of the Intern. School of Physics "Enrico Fermi"*, ed. by B. Bertotti, Academic Press, New York

Fairhead, L., Bretagnon, P., Lestrade, J.-F., 1986, in: *The Earth's Rotation and Reference Frames for Geodesy and Geodynamics*, G. Wilkins and A. Babcock (eds.), Reidel, Dordrecht

Finkelstein, A., Kreinovich, V., Pandey, S., 1983, *Astrophys. Space Sci.* **94**, 233

Fock, V.A., 1964, *The Theory of Space, Time and Gravitation*, Pergamon, New York

Fokker, A.D., 1921, *Proc. Roy. Acad. Amsterdam* **23**, 729

— 1965, *Time and Space, Weight and Inertia, A chronogeometrical introduction to Einstein's theory*, Pergamon, New York

Fomalont, E., Sramek, R., 1975, *Astrophys. J.* **199**, 749

— 1976, *Phys. Rev. Lett.* **36**, 1475

— 1977, *Comm. Astrophys.* **7**, 19

Fox, K., 1984, *Celest. Mech.* **33**, 127

Freundlich, E., v. Klüber, H., v. Brunn, A., 1931a, *Ab. Preuss. Akad. Wiss.* No. 1

— 1931b, *Z. Astrophys.* **3**, 171

Fricke, W., 1977, *Veröff. Astron. Rechen-Institut*, Heidelberg, No. 28

Fricke, W., 1981, in E. M. Gaposchkin and B. Kolaczek (eds.), *Reference Coordinate Systems for Earth Dynamics*, IAU Colloq. **56**, 331

Fujimoto, M.-K., Grafarend, E., 1986, in: *Relativity in Celestial Mechanics and Astrometry*, ed. by J. Kovalevsky and V. A. Brumberg, Reidel, Dordrecht

Fukushima, T., Fujimoto, M.-K., Kinoshita, H., Aoki, S., 1986, *Celest. Mech.* **38**, 215

Grishchuk, L., Kopejkin, S., 1986, in: *Relativity in Celestial Mechanics and Astrometry*, ed. by J. Kovalevsky and V. A. Brumberg, Reidel, Dordrecht

Guinot, B., 1986, *Celest. Mech.* **38**, 155

Haugan, M., 1985, *Astrophys. J.* **296**, 1

Havas, P., 1964, *Rev. Mod. Phys.* **36** , 938

Hawking, S. W., Ellis, G. F., 1973, *The Large Scale Structure of Space-Time*, Cambridge University Press, Cambridge

Hellings, R. W., Adams, P. J., Anderson, J. D., Keesey, M. S., Lau, E. L., Standish, E. M., Canuto, V. M., Goldman, I., 1983, *Phys. Rev. Lett.* **51**, 1609

Hellings, R. W., 1983, in: *10th Intern. Conf. on General Relativity and Gravitation, Conf. Papers*

— 1986, *Astron. J.* **91**, 650

Henrard, J., 1979, *Celest. Mech.* **19**, 337

Hill, H., 1971, in: *Proc. of the Conf. on Experimental Tests of Gravitation Theories*, ed. by R. Davies, p. 89, NASA-JPL Technical Memorandum 33 – 499

Hill, H. A., Stebbins, R. T., 1975, *Astrophys. J.* **200**, 471

Hill, H. A., Rabaey, G. R., Rosenwald, R. D., 1986, in: *Relativity in Celestial Mechanics and Astrometry*, ed. by J. Kovalevsky and V. A. Brumberg, Reidel, Dordrecht

Hirayama, T., Kinoshita, H., 1986, *Proc. of the 9th Symp. on Celestial Mechanics*, Akashi, Japan, 1987

Hirayama, T., Kinoshita, H., Fujimoto, M.-K., Fukushima, T., 1987, *Proc. of the IUGG General Assembly, Vancouver, August 1987*

Hogan, P. A., McCrea, J. D., 1974, *Gen. Rel. Grav.* **5**, 79

Hulse, R. A., Taylor, J. H., 1975, *Astrophys. J. Lett.* **195**, L51

Infeld, L., Plebanski, J., 1960, *Motion and Relativity*, Oxford University Press, Oxford

Jackson, J. D., 1975, *Classical Electrodynamics*, Wiley, New York

Jones, B., 1976, *Astron. J.* **81**, 455

Kalitzin, N. St., 1959, *Nuovo Cim.* **11**, 178

Kaplan, G. H., 1981, *The IAU Resolutions on Astronomical Constants, Time Scale, and the Fundamental Reference Frame*, U.S. Naval Observatory, Circular No. 163, Washington, D.C.

Kerlick, G. D., 1980, *Gen. Rel. Grav.* **12**, 467 and 521

Knowles, S., Wallmann, W., Hulburt, E., Cannon, W., Davidson, D., Patraschenko, W., Yen, J., Popelar, J., Galt, J., 1982, *NOAA Techn. Report* No. 95, NGS 24

Kopejkin, S. M., 1985, *Sov. Astron.* **29**, 516

Kovalevsky, J., 1984, *Space Sci. Rev.* **39**, 1

Kramer, D., Stephani, H., MacCallum, M., Herlt, E., 1980, *Exact Solutions of Einstein's Field Equations*, Cambridge University Press, Cambridge

Krogh, Ch., Baierlein, R., 1968, *Phys. Rev.* **175**, 1576

Künzle, H.P., 1976, *Gen. Rel. Grav.* **7**, 445

Labeyrie, A., 1975, *Astrophys. J. Letters* **196**, L71

Labeyrie, A., 1978, *Ann. Rev. Astron. Astrophys.* **16**, 77

Labeyrie, A., 1981, in *Proc. of the ESO Conf. on High Angular Resolution*, Garching, pp. 87 and 225

Landau, L. D., Lifschitz, E. M., 1962, *The Classical Theory of Fields*, Addison Wesley, Reading, Mass.

Lense, J., Thirring, H., 1918, *Phys. Z.* **19**, 156

Lestrade, J.-F., 1981, *Astron. Astrophys.* **100**, 143

Lestrade, J.-F., Bretagnon, P., 1982, *Astron. Astrophys.* **105**, 42

Levi-Civita, T., 1937, *Am. J. Math.* **59**, 9 and 225

— 1950, *Mémorial des Sciences Mathématiques* **116**, Gauthier-Villars, Paris

Li, W.-Q., Ni, W.-T., 1979, *J. Math. Phys.* **20**, 1473

Lipa, J. A., Fairbank, W. M., C. W. F., 1974, in: *Experimental Gravitation: Proc. of Course 56 of the Intern. School of Physics "Enrico Fermi"*, ed. by B. Bertotti, Academic Press, New York

Lipa, J. A., Everitt, C. W. F., 1978, *Acta Astronautica* **5**, 119

LMSC, 1982, *Comparative Feasibility Study of Two Concepts for a Space Based Astrometric Satellite*, Lockheed Palo Alto Research Laboratory document LMSC-D870885, August 1982

Lorentz, H. A., Droste, J., 1917, *Versl. K. Adad. Wet. Amsterdam* **26**, 392 and 649, reprinted in the Collected Papers of H.A. Lorentz, vol. 5, p. 330, The Hague, Nijhoff, 1937

Lottermoser, M., 1988, Ph.D. Thesis, Univ. of Munich

Manasse, F. K., Misner, C. W., 1963, *J. Math. Phys.* **4**, 735

Martin, C. F., Torrence, M. H., Misner, C. W., 1985, *J. Geophys. Res.* **90**, 9403

Mashhoon, B., 1985, *Found. Phys.* **15**, 497

Matukuma, T., Onuki, A., Yosida, S., Iwana, Y., 1940, *Jap. J. Astron. and Geophys.* **18**, 51

Merit Standards, 1983, *U.S. Naval Observatory Circular No.* **167**

Michalska, R., 1960, *Bull. Acad. Polon. Sci., Sér. sci. math. astr. et phys.* **8**, 247

Mikhailov, A., 1940, *C.R. Acad. Sci. USSR (N.S.)* **29**, 189

Misner, C. W., 1969, in: *Gravitational Collapse*, Brandeis Summer Institute 1968, M. Chrétien, S. Deser and J. Goldstein (eds.), Gordon and Breach, New York

Misner, C. W., Thorne, K. S., Wheeler, J. A., 1973, *Gravitation*, Freeman, San Francisco

Moyer, T. D., 1981, *Celest. Mech.* **23**, 33 and 57

Muhleman, D., Ekers, R., Fomalont, E., 1970, *Phys. Rev. Lett.* **24**, 1377

Murray, C. A., 1981, *Mon. Not. Roy. Ast. Soc.* **195**, 639

— 1983, *Vectorial Astrometry*, Adam Hilger, Bristol

Nordtvedt, K.,Jr., 1968, *Phys. Rev.* **D169**, 1014 and 1017

— 1970, *Icarus* **12**, 91

— 1971, *Phys. Rev.* **D3**, 1683

Ohta, T., Okamura, H., Kimura, T., Hiida, K., 1973, *Prog. Theor. Phys.* **50**, 492

— 1974, *Prog. Theor. Phys.* **51**, 1220 and 1598

Paik, H. J., Mashhoon, B., Will, C., 1988, *Detection of gravitomagnetic field using an orbiting superconducting gravity gradiometer*, preprint

Papapetrou, A., 1951, *Proc. Phys. Soc.* **A64**, 57

Pavlov, N. V., 1984, *Sov. Astron.* **28**, 223 and 351

Peres, A., 1959a, *Nuovo Cim.* **11**, 617 and 644

— 1959b, *Nuovo Cim.* **13**, 437

— 1960, *Nuovo Cim.* **15**, 351

Peters, P. C., Mathews, J., 1963, *Phys. Rev.* **131**, 435

Plummer, H. C., 1960, *An Introductory Treatise on Dynamical Astronomy*, Dover, New York

Pugh, G. E., 1959, *WSEG Research Memo* **11**, U.S. Dept. of Defence

Reasenberg, R. D., Shapiro, I. I., 1976, in: *Atomic Masses and Fundamental Constants*, Vol.5, ed. by J. H. Sanders and A. H. Wapstra, Plenum Press, New York

Reasenberg, R. D., Shapiro, I. I., 1978, in: *On the Measurement of Cosmological Variations of the Gravitational Constant*, ed. L. Halpern, 71, University Presses of Florida, Gainesville

Reasenberg, R. D., Shapiro, I. I., MacNeil, P. E., Goldstein, R. B., Breidenthal, J. C., Brenkle, J. P., Cain, D. L., Kaufman, T. M., Komarek, T. A., Zygielbaum, A. I., 1979, *Astrophys. J.* **234**, L219

Reasenberg, R., Shapiro, I., 1982, *Acta Astronautica* **9**, 103

Riley, J., 1973, *Mon. Not. Roy. Ast. Soc.* **161**, 11P

Robertson, D. S., Carter, W. E., 1984, *Nature* **310**, 572

Rosenblum, A., 1978, *Phys. Rev. Lett.* **41**, 1003

— 1981, *Phys. Lett.* **81A**, 1

Roseveare, N. T., 1982, *Mercury's Perihelion from Le Verrier to Einstein*, Clarendon Press, Oxford

Rotge, J., Shaw, G., Emrick, H., 1985, in: *Proc. of the Intern. Conf. on Earth Rotation and the Terrestrial Reference Frame*, July 31 – Aug. 2, Columbus, Ohio, Vol.II, 719 ed. by I. Mueller

Roy, A. E., 1978, *Orbital Motion*, Adam Hilger, Bristol

Rubincam, D. P., 1982, *Celest. Mech.* **26**, 361

Sachs, R. K., Wu, H., 1977, *General Relativity for Mathematicians*, Springer, Berlin, Heidelberg, New York

Schäfer, G., 1982, *Prog. Theor. Phys.* **68**, 2191

— 1983, *Lett. Nuovo Cim.* **36**, 105

— 1985, *Ann. Phys. (NY)* **161**, 81

— 1987, *Phys. Lett.* **A123**, 336

Schattner, R., Lasitzky, G., 1984, *Ann. Inst. Henri Poincaré* **40**, 291

Schastok, J., Gleixner, H., Soffel, M. H., Ruder, H., Schneider, M., 1988, submitted to Comp. Phys. Comm.

Schiff, L. I., 1960, *Phys. Rev. Lett.* **4**, 215

Schmeidler, F., 1962, *Astron. Nachr.* **287**, 7

— 1984, private communication

Scully, M. O., Zubairy, M. S., Haugan, M. P., 1981, *Phys. Rev.* **A24**, 2009

Seielstad, G., Sramek, R., Weiler, K., 1970, *Phys. Rev. Lett.* **24**, 1373

Shapiro, I. I., 1964, *Phys. Rev. Lett.* **13**, 789

Shapiro, I. I., Counselman, C. C., King, R. W., 1976, *Phys. Rev. Lett.* **36**, 555

Smarr, L. L., Blandford, R., 1976, *Astrophys. J.* **207**, 574

Smith, D. E., Dunn, P., 1980, *Geophys. Res. Lett.* **7**, 437

Soffel, M. H., Schastok, J., Ruder, H., Schneider, M., 1985, *Astrophys. Space Sci.* **110**, 95

Soffel, M. H., Ruder, H., Schneider, M., 1986, *Astron. Astrophys.* **157**, 357

Soffel, M. H., Herold, H., Ruder, H., Schneider, M., 1988a, *Manuscripta Geodaetica* **13**, 139

Soffel, M. H., Wirrer, R., Schastok, J., Ruder, H., Schneider, M., 1988b, *Relativistic Effects in the Motion of Artificial Satellites: The Oblateness of the Central Body*, *Celest. Mech.* in press

Spyrou, N., 1978, *Gen. Rel. Grav.* **9**, 519

Sramek, R. A., 1971, *Astrophys. J.* **167**, L55

Sramek, R. A., 1974, in: *Experimental Gravitation: Proceedings of Course 56 of the International School of Physics "Enrico Fermi"*, ed. by B. Bertotti, 163, Academic, New York

Straumann, N., 1984, *General Relativity and Relativistic Astrophysics*, Springer, Berlin, Heidelberg, New York

Stumpff, P., 1979, *Astron. Astrophys.* **78**, 229

Synge, J. L., 1966, *Relativity: The General Theory*, North-Holland, Amsterdam

— 1969, *Proc. Roy. Ir. Acad.* **A67**, 47

— 1970, *Proc. Roy. Ir. Acad.* **A69**, 11

Taylor, J. H., Weisberg, J. M., 1982, *Astrophys. J.* **253**, 908

Taylor, J. H., 1987, in: *General Relativity and Gravitation*, ed. by M. A. MacCallum, Cambridge University Press, Cambridge

Texas Mauritanian Eclipse Team, 1976, *Astron. J.* **81**, 452

Thirring, H., 1918, *Phys. Z.* **19**, 33

Thomas, J. B., 1975, *Astron. J.* **80**, 405

Thomas, L. H., 1927, *Phil. Mag.* **3**, 1

Thorne, S. K., 1980, *Rev. Mod. Phys.* **52**, 285

Torge, W., 1975, *Geodäsie*, Walter de Gruyter, Berlin

Trautman, A., 1965, in: *Lectures on General Relativity*, Brandeis 1964 Summer Institute on Theoretical Physics, vol. I, Prentice Hall, Englewood Cliffs N.J., ed. by A. Trautmann, F. Pirani and H. Bondi

van Biesbroeck, G., 1949, *Astron. J.* **55**, 49 and 247

— 1953, *Astron. J.* **58**, 87

van de Hulst, H. C., 1950, *Bull. Astron. Inst. Neth.* **11**, 135

Vessot, R. F. C., Levine, M. W., 1979, *Gen. Rel. and Grav.* **10**, 181

Vessot, R. F. C., Levine, M. W., Mattison, E. M., Blomberg, E. L., Hoffmann, T. E., Nystrom, G. U., Farrel, B. F., Decher, R., Eby, P. B., Baugher, C. R., Watts, J. W., Teuber, D. L., Wills, F. O., 1980, *Phys. Rev. Lett.* **45**, 2081

Vessot, R. F. C., 1984, *Contemp. Phys.* **25**, 355

Vincent, M. A., 1984, *The Determination of the Post-Newtonian Parameters in Gravitational Theory Using Laser Ranging to the LAGEOS Satellite*, Ph. D. Thesis, Univ. of Texas, Austin

— 1986, *Celest. Mech.* **39**, 15

von Klüber, H., 1960, in: *Vistas in Astronomy*, ed. by A. Beer, Pergamon Press, New York

Wagoner, R., Will, C., 1976, *Astrophys. J.* **210**, 764

Walker, M., Will, C., 1980, *Phys. Rev. Lett.* **45**, 1741

Weiler, K., Ekers, R., Raimond, E., Wellington, K., 1975, *Phys. Rev. Lett.* **35**, 134

Weinberg, S., 1972, *Gravitation and Cosmology*, Wiley, New York

Weisberg, J. M., Taylor, J. H., 1984, *Phys. Rev. Lett.* **52**, 1348

Westpfahl, K., Möhles, R., Simonis, H., 1987, *Class. Quantum Grav.* **4**, L185

Will, C., 1971, *Astrophys. J.* **169**, 141

Will, C., 1981, *Theory and Experiment in Gravitational Physics*, Cambridge University Press, Cambridge

Will, C., 1984, *Phys. Rep.* **113**, 345

Williams, J. G., Dicke, R. H., Bender, P. L., Alley, C. O., Carter, W. E., Currie, D. G., Eckhardt, D. H., Faller, J. E., Kaula, W. M., Mulholland, J. D., Plotkin, H. H., Poultney, S. K., Shelus, P. J., Silverberg, E. C., Sinclair, W. S., Slade, M. A., Wilkinson, D. T., 1976, *Phys. Rev. Lett.* **36**, 551

Worden, P. W., Jr., 1978, *Acta Astronautica* **5**, 27

Worden, P. W., Jr., C. W. F., Everitt, 1974, in: *Experimental Gravitation: Proc. of Course 56 of the Intern. School of Physics "Enrico Fermi"*, ed. by B. Bertotti, Academic Press, New York

Worden, S., Lynds, C., Harvey, J., 1976, *J. Opt. Soc. Am.* **66**, 1243

Subject Index

Note Added in Proof

After the completion of this book important progress has been achieved on the subject under discussion. Brumberg and Kopejkin (1988,1989) have constructed a "good" local (GRS) coordinate system comoving with the Earth that contains the tidal forces, extending our results from Sections 3.4 and 3.5. Their GRS-coordinates are harmonic and have been determined by asymptotic matching of the (local) GRS-metric with the global barycentric (BRS-) metric in the multipole formalism (see also Voinov 1988).

Shapiro et al. (1988) have experimentally determined the geodetic (de Sitter) precession in the Moon's orbit from lunar laser-ranging data. Their result confirms the prediction from general relativity ($\sim 2''$ per century) within an estimated error of $0.04''/\mathrm{cy}$.

In Nordtvedt (1988) it is demonstrated how the existence of the gravito-magnetic interaction can be inferred from laser ranging to the LAGEOS satellite.

References:

Brumberg, V.A., Kopejkin, S.M., 1988, in: *Reference Systems* (eds. J.Kovalevsky, I.Mueller and B.Kolaczek), Reidel, Dordrecht

Brumberg, V.A., Kopejkin, S.M., 1989, *Nuovo Cim. B*, in press

Nordtvedt, K., 1988, *Phys. Rev. Lett.* **61**, 2647

Shapiro, I.I., Reasenberg, R.D., Chandler, J.F., Babcock, R.W., 1988, *Phys. Rev. Lett.* **61**, 2643

Voinov, A.V., 1988, *Celest. Mech.* **42**, 293

Errata

p. 129: the second term on the r.h.s. of (4.7.47) reads

$$\int \frac{\rho^* \rho^{*\prime}(x^i - x^{i\prime})(x^j - x^{j\prime})}{|\mathbf{x} - \mathbf{x}'|^3} \, d^3x \, d^3x'$$

p. 192: in the second equation the middle term reads

$$(\rho v^i v^j + p\delta_{ij})_{,j}$$

p. 192: in A4.4 the divergence comprises the $p\,\delta_{ij}$-term

p. 193: the last term in f^i reads

$$\frac{1}{2}\rho^* v^j (V^i - W^i)_{,j}.$$